California Algebra 1

Study Guide and Intervention Workbook

New York, New York · Columbus, Ohio · Chicago, Illinois · Woodland Hills, California

TO THE STUDENT This *Study Guide and Intervention Workbook* gives you additional examples and problems for the concept exercises in each lesson. The exercises are designed to aid your study of mathematics by reinforcing important mathematical skills needed to succeed in the everyday world. The materials are organized by chapter and lesson, with two *Study Guide and Intervention* worksheets for every lesson in *Glencoe Algebra 1*.

Always keep your workbook handy. Along with your textbook, daily homework, and class notes, the completed *Study Guide and Intervention Workbook* can help you review for quizzes and tests.

TO THE TEACHER These worksheets are the same as those found in the Chapter Resource Masters for *Glencoe Algebra 1*. The answers to these worksheets are available at the end of each Chapter Resource Masters booklet as well as in your Teacher Wraparound Edition interleaf pages.

Copyright © by The McGraw-Hill Companies, Inc. All rights reserved. Except as permitted under the United States Copyright Act, no part of this publication may be reproduced or distributed in any form or by any means, or stored in a database or retrieval system, without prior written permission of the publisher.

Send all inquiries to:
Glencoe/McGraw-Hill
8787 Orion Place
Columbus, OH 43240

ISBN: 978-0-07-879045-4
MHID: 0-07-879045-X

Study Guide And Intervention Workbook, Algebra 1

Printed in the United States of America

1 2 3 4 5 6 7 8 9 10 009 14 13 12 11 10 09 08 07

CONTENTS

Lesson/Title	Page
1-1 Variables and Expressions	1
1-2 Order of Operations	3
1-3 Open Sentences	5
1-4 Identity and Equality Properties	7
1-5 The Distributive Property	9
1-6 Commutative and Associative Properties	11
1-7 Logical Reasoning and Counterexamples	13
1-8 Number Systems	15
1-9 Functions and Graphs	17
2-1 Writing Equations	19
2-2 Solving Equations by Using Addition and Subtraction	21
2-3 Solving Equations by Using Multiplication and Division	23
2-4 Solving Multi-Step Equations	25
2-5 Solving Equations with the Variable on Each Side	27
2-6 Ratios and Proportions	29
2-7 Percent of Change	31
2-8 Solving Equations and Formulas	33
2-9 Weighted Averages	35
3-1 Representing Relations	37
3-2 Representing Functions	39
3-3 Linear Functions	41
3-4 Arithmetic Sequences	43
3-5 Describing Number Patterns	45
4-1 Rate of Change and Slope	47
4-2 Slope and Direct Variation	49
4-3 Graphing Equations in Slope-Intercept Form	51
4-4 Writing Equations in Slope-Intercept Form	53
4-5 Writing Equations in Point-Slope Form	55
4-6 Statistics: Scatter Plots and Lines of Fit	57
4-7 Geometry: Parallel and Perpendicular Lines	59
5-1 Graphing Systems of Equations	61
5-2 Substitution	63
5-3 Elimination Using Addition and Subtraction	65
5-4 Elimination Using Multiplication	67

Lesson/Title	Page
5-5 Applying Systems of Linear Equations	69
6-1 Solving Inequalities by Addition and Subtraction	71
6-2 Solving Inequalities by Multiplication and Division	73
6-3 Solving Multi-Step Inequalities	75
6-4 Solving Compound Inequalities	77
6-5 Solving Open Sentences Involving Absolute Value	79
6-6 Solving Inequalities Involving Absolute Value	81
6-7 Graphing Inequalities in Two Variables	83
6-8 Graphing Systems of Inequalities	85
7-1 Multiplying Monomials	87
7-2 Dividing Monomials	89
7-3 Polynomials	91
7-4 Adding and Subtracting Polynomials	93
7-5 Multiplying a Polynomial by a Monomial	95
7-6 Multiplying Polynomials	97
7-7 Special Products	99
8-1 Monomials and Factoring	101
8-2 Factoring Using the Distributive Property	103
8-3 Factoring Trinomials: $x^2 + bx + c$	105
8-4 Factoring Trinomials: $ax^2 + bx + c$	107
8-5 Factoring Differences of Squares	109
8-6 Perfect Squares and Factoring	111
9-1 Graphing Quadratic Functions	113
9-2 Solving Quadratic Equations by Graphing	115
9-3 Solving Quadratic Equations by Completing the Square	117
9-4 Solving Quadratic Equations by Using the Quadratic Formula	119
9-5 Exponential Functions	121
9-6 Growth and Decay	123
10-1 Simplifying Radical Expressions	125
10-2 Operations with Radical Expressions	127
10-3 Radical Equations	129
10-4 The Pythagorean Theorem	131
10-5 The Distance Formula	133
10-6 Similar Triangles	135

Lesson/Title		Page
11-1	Inverse Variation	137
11-2	Rational Expressions	139
11-3	Multiplying Rational Expressions	141
11-4	Dividing Rational Expressions	143
11-5	Dividing Polynomials	145
11-6	Rational Expressions with Like Denominators	147
11-7	Rational Expressions with Unlike Denominators	149
11-8	Mixed Expressions and Complex Fractions	151
11-9	Solving Rational Equations	153
12-1	Sampling and Bias	155
12-2	Counting Outcomes	157
12-3	Permutations and Combinations	159
12-4	Probability of Compound Events	161
12-5	Probability Distributions	163
12-6	Probability Simulations	165

NAME _____ DATE _____ PERIOD _____

1-1 Study Guide and Intervention

Variables and Expressions

Write Mathematical Expressions In the **algebraic expression**, ℓw, the letters ℓ and w are called **variables**. In algebra, a variable is used to represent unspecified numbers or values. Any letter can be used as a variable. The letters ℓ and w are used above because they are the first letters of the words *length* and *width*. In the expression ℓw, ℓ and w are called factors, and the result is called the **product**.

Example 1 Write an algebraic expression for each verbal expression.

a. four more than a number n
  The words *more than* imply addition.
  four more than a number n
  $4 + n$
  The algebraic expression is $4 + n$.

b. the difference of a number squared and 8
  The expression *difference of* implies subtraction.
  the difference of a number squared and 8
  $n^2 - 8$
  The algebraic expression is $n^2 - 8$.

Example 2 Evaluate each expression.

a. 3^4
  $3^4 = 3 \cdot 3 \cdot 3 \cdot 3$ Use 3 as a factor 4 times.
  $= 81$ Multiply.

b. five cubed
  Cubed means raised to the third power.
  $5^3 = 5 \cdot 5 \cdot 5$ Use 5 as a factor 3 times.
  $= 125$ Multiply.

Exercises

Write an algebraic expression for each verbal expression.

1. a number decreased by 8

2. a number divided by 8

3. a number squared

4. four times a number

5. a number divided by 6

6. a number multiplied by 37

7. the sum of 9 and a number

8. 3 less than 5 times a number

9. twice the sum of 15 and a number

10. one-half the square of b

11. 7 more than the product of 6 and a number

12. 30 increased by 3 times the square of a number

Evaluate each expression.

13. 5^2

14. 3^3

15. 10^4

16. 12^2

17. 8^3

18. 2^8

Study Guide and Intervention

Glencoe Algebra 1

1-1 Study Guide and Intervention (continued)

Variables and Expressions

Write Verbal Expressions Translating algebraic expressions into verbal expressions is important in algebra.

Example Write a verbal expression for each algebraic expression.

a. $6n^2$

the product of 6 and n squared

b. $n^3 - 12m$

the difference of n cubed and twelve times m

Exercises

Write a verbal expression for each algebraic expression.

1. $w - 1$

2. $\frac{1}{3}a^3$

3. $81 + 2x$

4. $12c$

5. 8^4

6. 6^2

7. $2n^2 + 4$

8. $a^3 \cdot b^3$

9. $2x^3 - 3$

10. $\frac{6k^3}{5}$

11. $\frac{1}{4}b^2$

12. $7n^5$

13. $3x + 4$

14. $\frac{2}{3}k^5$

15. $3b^2 + 2a^3$

16. $4(n^2 + 1)$

17. $3^2 + 2^3$

18. $6n^2 + 3$

NAME _____ DATE _____ PERIOD _____

1-2 Study Guide and Intervention

Order of Operations

Evaluate Rational Expressions Numerical expressions often contain more than one operation. To evaluate them, use the rules for order of operations shown below.

Order of Operations	Step 1 Evaluate expressions inside grouping symbols.
	Step 2 Evaluate all powers.
	Step 3 Do all multiplication and/or division from left to right.
	Step 4 Do all addition and/or subtraction from left to right.

Example 1 Evaluate each expression.

a. $7 + 2 \cdot 4 - 4$

$7 + 2 \cdot 4 - 4 = 7 + 8 - 4$ Multiply 2 and 4.
$= 15 - 4$ Add 7 and 8.
$= 11$ Subtract 4 from 15.

b. $3(2) + 4(2 + 6)$

$3(2) + 4(2 + 6) = 3(2) + 4(8)$ Add 2 and 6.
$= 6 + 32$ Multiply left to right.
$= 38$ Add 6 and 32.

Example 2 Evaluate each expression.

a. $3[2 + (12 \div 3)^2]$

$3[2 + (12 \div 3)^2] = 3(2 + 4^2)$ Divide 12 by 3.
$= 3(2 + 16)$ Find 4 squared.
$= 3(18)$ Add 2 and 16.
$= 54$ Multiply 3 and 18.

b. $\dfrac{3 + 2^3}{4^2 \cdot 3}$

$\dfrac{3 + 2^3}{4^2 \cdot 3} = \dfrac{3 + 8}{4^2 \cdot 3}$ Evaluate power in numerator.

$= \dfrac{11}{4^2 \cdot 3}$ Add 3 and 8 in the numerator.

$= \dfrac{11}{16 \cdot 3}$ Evaluate power in denominator.

$= \dfrac{11}{48}$ Multiply.

Exercises

Evaluate each expression.

1. $(8 - 4) \cdot 2$

2. $(12 + 4) \cdot 6$

3. $10 + 2 \cdot 3$

4. $10 + 8 \cdot 1$

5. $15 - 12 \div 4$

6. $\dfrac{15 + 60}{30 - 5}$

7. $12(20 - 17) - 3 \cdot 6$

8. $24 \div 3 \cdot 2 - 3^2$

9. $8^2 \div (2 \cdot 8) + 2$

10. $3^2 \div 3 + 2^2 \cdot 7 - 20 \div 5$

11. $\dfrac{4 + 3^2}{12 + 1}$

12. $\dfrac{8(2) - 4}{8 \div 4}$

13. $250 \div [5(3 \cdot 7 + 4)]$

14. $\dfrac{2 \cdot 4^2 - 8 \div 2}{(5 + 2) \cdot 2}$

15. $\dfrac{4 \cdot 3^2 - 3 \cdot 2}{3 \cdot 5}$

16. $\dfrac{4(5^2) - 4 \cdot 3}{4(4 \cdot 5 + 2)}$

17. $\dfrac{5^2 - 3}{20(3) + 2(3)}$

18. $\dfrac{8^2 - 2^2}{(2 \cdot 8) + 4}$

Study Guide and Intervention   Glencoe Algebra 1

1-2 Study Guide and Intervention (continued)
Order of Operations

Evaluate Algebraic Expressions Algebraic expressions may contain more than one operation. Algebraic expressions can be evaluated if the values of the variables are known. First, replace the variables with their values. Then use the order of operations to calculate the value of the resulting numerical expression.

Example Evaluate $x^3 + 5(y - 3)$ if $x = 2$ and $y = 12$.

$x^3 + 5(y - 3) = 2^3 + 5(12 - 3)$ Replace x with 2 and y with 12.
$ = 8 + 5(12 - 3)$ Evaluate 2^3.
$ = 8 + 5(9)$ Subtract 3 from 12.
$ = 8 + 45$ Multiply 5 and 9.
$ = 53$ Add 8 and 45.

The solution is 53.

Exercises

Evaluate each expression if $x = 2$, $y = 3$, $z = 4$, $a = \frac{4}{5}$, and $b = \frac{3}{5}$.

1. $x + 7$

2. $3x - 5$

3. $x + y^2$

4. $x^3 + y + z^2$

5. $6a + 8b$

6. $23 - (a + b)$

7. $\dfrac{y^2}{x^2}$

8. $2xyz + 5$

9. $x(2y + 3z)$

10. $(10x)^2 + 100a$

11. $\dfrac{3xy - 4}{7x}$

12. $a^2 + 2b$

13. $\dfrac{z^2 - y^2}{x^2}$

14. $6xz + 5xy$

15. $\dfrac{(z - y)^2}{x}$

16. $\dfrac{25ab + y}{xz}$

17. $\dfrac{5a^2 b}{y}$

18. $(z \div x)^2 + ax$

19. $\left(\dfrac{x}{z}\right)^2 + \left(\dfrac{y}{z}\right)^2$

20. $\dfrac{x + z}{y + 2z}$

21. $\left(\dfrac{z \div x}{y}\right) + \left(\dfrac{y \div x}{z}\right)$

NAME _____ DATE _____ PERIOD _____

1-3 Study Guide and Intervention

Open Sentences

Solve Equations A mathematical sentence with one or more variables is called an **open sentence**. Open sentences are **solved** by finding replacements for the variables that result in true sentences. The set of numbers from which replacements for a variable may be chosen is called the **replacement set**. The set of all replacements for the variable that result in true statements is called the **solution set** for the variable. A sentence that contains an equal sign, =, is called an **equation**.

Example 1 Find the solution set of $3a + 12 = 39$ if the replacement set is {6, 7, 8, 9, 10}.

Replace a in $3a + 12 = 39$ with each value in the replacement set.

$3(6) + 12 \stackrel{?}{=} 39 \to 30 \neq 39$ false
$3(7) + 12 \stackrel{?}{=} 39 \to 33 \neq 39$ false
$3(8) + 12 \stackrel{?}{=} 39 \to 36 \neq 39$ false
$3(9) + 12 \stackrel{?}{=} 39 \to 39 = 39$ true
$3(10) + 12 \stackrel{?}{=} 39 \to 42 \neq 39$ false

Since $a = 9$ makes the equation $3a + 12 = 39$ true, the solution is 9. The solution set is {9}.

Example 2 Solve $\frac{2(3+1)}{3(7-4)} = b$.

$\frac{2(3+1)}{3(7-4)} = b$ Original equation

$\frac{2(4)}{3(3)} = b$ Add in the numerator; subtract in the denominator.

$\frac{8}{9} = b$ Simplify.

The solution is $\frac{8}{9}$.

Exercises

Find the solution of each equation if the replacement sets are $X = \left\{\frac{1}{4}, \frac{1}{2}, 1, 2, 3\right\}$ and $Y = \{2, 4, 6, 8\}$.

1. $x + \frac{1}{2} = \frac{5}{2}$

2. $x + 8 = 11$

3. $y - 2 = 6$

4. $x^2 - 1 = 8$

5. $y^2 - 2 = 34$

6. $x^2 + 5 = 5\frac{1}{16}$

7. $2(x + 3) = 7$

8. $\frac{1}{4}(y + 1)^2 = \frac{9}{4}$

9. $y^2 + y = 20$

Solve each equation.

10. $a = 2^3 - 1$

11. $n = 6^2 - 4^2$

12. $w = 6^2 \cdot 3^2$

13. $\frac{1}{4} + \frac{5}{8} = k$

14. $\frac{18 - 3}{2 + 3} = p$

15. $s = \frac{15 - 6}{27 - 24}$

16. $18.4 - 3.2 = m$

17. $k = 9.8 + 5.7$

18. $c = 3\frac{1}{2} + 2\frac{1}{4}$

NAME _____ DATE _____ PERIOD _____

1-3 Study Guide and Intervention *(continued)*

Open Sentences

Solve Inequalities An open sentence that contains the symbol $<$, $\leq$, $>$, or $\geq$ is called an **inequality**. Inequalities can be solved the same way that equations are solved.

Example Find the solution set for $3a - 8 > 10$ if the replacement set is $\{4, 5, 6, 7, 8\}$.

Replace a in $3a - 8 > 10$ with each value in the replacement set.

$3(4) - 8 \stackrel{?}{>} 10 \rightarrow 4 \not> 10$ false
$3(5) - 8 \stackrel{?}{>} 10 \rightarrow 7 \not> 10$ false
$3(6) - 8 \stackrel{?}{>} 10 \rightarrow 10 \not> 10$ false
$3(7) - 8 \stackrel{?}{>} 10 \rightarrow 13 > 10$ true
$3(8) - 8 \stackrel{?}{>} 10 \rightarrow 16 > 10$ true

Since replacing a with 7 or 8 makes the inequality $3a - 8 > 10$ true, the solution set is $\{7, 8\}$.

Exercises

Find the solution set for each inequality if the replacement set is $X = \{0, 1, 2, 3, 4, 5, 6, 7\}$.

1. $x + 2 > 4$
2. $x \div 3 < 6$
3. $3x > 18$

4. $\dfrac{x}{3} > 1$
5. $\dfrac{x}{5} \geq 2$
6. $\dfrac{3x}{8} \leq 2$

7. $3x - 4 > 5$
8. $3(8 - x) + 1 \leq 6$
9. $4(x + 3) \geq 20$

Find the solution set for each inequality if the replacement sets are $X = \left\{\dfrac{1}{4}, \dfrac{1}{2}, 1, 2, 3, 5, 8\right\}$ and $Y = \{2, 4, 6, 8, 10\}$.

10. $x + 3 > 5$
11. $y \div 3 < 6$
12. $8y + 3 \geq 51$

13. $\dfrac{x}{2} < 4$
14. $\dfrac{y}{4} \geq 2$
15. $\dfrac{2y}{5} \leq 2$

16. $4x + 1 \geq 4$
17. $3x + 3 \geq 12$
18. $2(y + 1) \geq 18$

19. $3x - \dfrac{1}{4} < 2$
20. $3y + 2 \leq 8$
21. $\dfrac{1}{2}(6 - 2x) + 2 \leq 3$

1-4 Study Guide and Intervention

Identity and Equality Properties

Identity and Equality Properties The identity and equality properties in the chart below can help you solve algebraic equations and evaluate mathematical expressions.

Additive Identity	For any number a, $a + 0 = a$.
Multiplicative Identity	For any number a, $a \cdot 1 = a$.
Multiplicative Property of 0	For any number a, $a \cdot 0 = 0$.
Multiplicative Inverse Property	For every number $\frac{a}{b}$, $a, b \neq 0$, there is exactly one number $\frac{b}{a}$ such that $\frac{a}{b} \cdot \frac{b}{a} = 1$.
Reflexive Property	For any number a, $a = a$.
Symmetric Property	For any numbers a and b, if $a = b$, then $b = a$.
Transitive Property	For any numbers a, b, and c, if $a = b$ and $b = c$, then $a = c$.
Substitution Property	If $a = b$, then a may be replaced by b in any expression.

Example 1 Name the property used in each equation. Then find the value of n.

a. $8n = 8$
Multiplicative Identity Property
$n = 1$, since $8 \cdot 1 = 8$

b. $n \cdot 3 = 1$
Multiplicative Inverse Property
$n = \frac{1}{3}$, since $\frac{1}{3} \cdot 3 = 1$

Example 2 Name the property used to justify each statement.

a $5 + 4 = 5 + 4$
Reflexive Property

b. If $n = 12$, then $4n = 4 \cdot 12$.
Substitution Property

Exercises

Name the property used in each equation. Then find the value of n.

1. $6n = 6$
2. $n \cdot 1 = 8$
3. $6 \cdot n = 6 \cdot 9$
4. $9 = n + 9$
5. $n + 0 = \frac{3}{8}$
6. $\frac{3}{4} \cdot n = 1$

Name the property used in each equation.

7. If $4 + 5 = 9$, then $9 = 4 + 5$.
8. $0 + 21 = 21$
9. $0(15) = 0$
10. $(1)94 = 94$
11. If $3 + 3 = 6$ and $6 = 3 \cdot 2$, then $3 + 3 = 3 \cdot 2$.
12. $4 + 3 = 4 + 3$
13. $(14 - 6) + 3 = 8 + 3$

1-4 Study Guide and Intervention (continued)

Identity and Equality Properties

Use Identity and Equality Properties The properties of identity and equality can be used to justify each step when evaluating an expression.

Example Evaluate $24 \cdot 1 - 8 + 5(9 \div 3 - 3)$. Name the property used in each step.

$24 \cdot 1 - 8 + 5(9 \div 3 - 3) = 24 \cdot 1 - 8 + 5(3 - 3)$	Substitution; $9 \div 3 = 3$
$= 24 \cdot 1 - 8 + 5(0)$	Substitution; $3 - 3 = 0$
$= 24 - 8 + 5(0)$	Multiplicative Identity; $24 \cdot 1 = 24$
$= 24 - 8 + 0$	Multiplicative Property of Zero; $5(0) = 0$
$= 16 + 0$	Substitution; $24 - 8 = 16$
$= 16$	Additive Identity; $16 + 0 = 16$

Exercises

Evaluate each expression. Name the property used in each step.

1. $2\left[\dfrac{1}{4} + \left(\dfrac{1}{2}\right)^2\right]$

2. $15 \cdot 1 - 9 + 2(15 \div 3 - 5)$

3. $2(3 \cdot 5 \cdot 1 - 14) - 4 \cdot \dfrac{1}{4}$

4. $18 \cdot 1 - 3 \cdot 2 + 2(6 \div 3 - 2)$

5. $10 \div 5 - 2^2 \div 2 + 13$

6. $3(5 - 5 \cdot 1^2) + 21 \div 7$

1-5 Study Guide and Intervention

The Distributive Property

Evaluate Expressions The Distributive Property can be used to help evaluate expressions.

Distributive Property	For any numbers a, b, and c, $a(b + c) = ab + ac$ and $(b + c)a = ba + ca$ and $a(b - c) = ab - ac$ and $(b - c)a = ba - ca$.

Example 1 Rewrite $6(8 + 10)$ using the Distributive Property. Then evaluate.

$6(8 + 10) = 6 \cdot 8 + 6 \cdot 10$ Distributive Property
$ = 48 + 60$ Multiply.
$ = 108$ Add.

Example 2 Rewrite $-2(3x^2 + 5x + 1)$ using the Distributive Property. Then simplify.

$-2(3x^2 + 5x + 1) = -2(3x^2) + (-2)(5x) + (-2)(1)$ Distributive Property
$ = -6x^2 + (-10x) + (-2)$ Multiply.
$ = -6x^2 - 10x - 2$ Simplify.

Exercises

Rewrite each expression using the Distributive Property. Then simplify.

1. $2(10 - 5)$
2. $6(12 - t)$
3. $3(x - 1)$

4. $6(12 + 5)$
5. $(x - 4)3$
6. $-2(x + 3)$

7. $5(4x - 9)$
8. $3(8 - 2x)$
9. $12\left(6 - \frac{1}{2}x\right)$

10. $12\left(2 + \frac{1}{2}x\right)$
11. $\frac{1}{4}(12 - 4t)$
12. $3(2x - y)$

13. $2(3x + 2y - z)$
14. $(x - 2)y$
15. $2(3a - 2b + c)$

16. $\frac{1}{4}(16x - 12y + 4z)$
17. $(2 - 3x + x^2)3$
18. $-2(2x^2 + 3x + 1)$

Study Guide and Intervention Glencoe Algebra 1

1-5 Study Guide and Intervention (continued)
The Distributive Property

Simplify Expressions A **term** is a number, a variable, or a product or quotient of numbers and variables. **Like terms** are terms that contain the same variables, with corresponding variables having the same powers. The Distributive Property and properties of equalities can be used to simplify expressions. An expression is in **simplest form** if it is replaced by an **equivalent** expression with no like terms or parentheses.

Example
Simplify $4(a^2 + 3ab) - ab$.

$4(a^2 + 3ab) - ab = 4(a^2 + 3ab) - 1ab$ Multiplicative Identity
$ = 4a^2 + 12ab - 1ab$ Distributive Property
$ = 4a^2 + (12 - 1)ab$ Distributive Property
$ = 4a^2 + 11ab$ Substitution

Exercises
Simplify each expression. If not possible, write *simplified*.

1. $12a - a$

2. $3x + 6x$

3. $3x - 1$

4. $12g - 10g + 1$

5. $-2x - 12$

6. $4x^2 + 3x + 7$

7. $20a + 12a - 8$

8. $3x^2 + 2x^2$

9. $-6x + 3x^2 + 10x^2$

10. $2p + \frac{1}{2}q$

11. $10xy - 4(xy + xy)$

12. $21c + 18c + 31b - 3b$

13. $3x - 2x - 2y + 2y$

14. $xy - 2xy$

15. $12a - 12b + 12c$

16. $4x + \frac{1}{4}(16x - 20y)$

17. $2 - 1 - 6x + x^2$

18. $4x^2 + 3x^2 + 2x$

NAME _____ DATE _____ PERIOD _____

1-6 Study Guide and Intervention

Commutative and Associative Properties

Commutative and Associative Properties The Commutative and Associative Properties can be used to simplify expressions. The Commutative Properties state that the order in which you add or multiply numbers does not change their sum or product. The Associative Properties state that the way you group three or more numbers when adding or multiplying does not change their sum or product.

Commutative Properties	For any numbers a and b, $a + b = b + a$ and $a \cdot b = b \cdot a$.
Associative Properties	For any numbers a, b, and c, $(a + b) + c = a + (b + c)$ and $(ab)c = a(bc)$.

Example 1 Evaluate $6 \cdot 2 \cdot 3 \cdot 5$.

$6 \cdot 2 \cdot 3 \cdot 5 = 6 \cdot 3 \cdot 2 \cdot 5$ Commutative Property
$= (6 \cdot 3)(2 \cdot 5)$ Associative Property
$= 18 \cdot 10$ Multiply.
$= 180$ Multiply.

The product is 180.

Example 2 Evaluate $8.2 + 2.5 + 2.5 + 1.8$.

$8.2 + 2.5 + 2.5 + 1.8$
$= 8.2 + 1.8 + 2.5 + 2.5$ Commutative Prop.
$= (8.2 + 1.8) + (2.5 + 2.5)$ Associative Prop.
$= 10 + 5$ Add.
$= 15$ Add.

The sum is 15.

Exercises

Evaluate each expression.

1. $12 + 10 + 8 + 5$

2. $16 + 8 + 22 + 12$

3. $10 \cdot 7 \cdot 2.5$

4. $4 \cdot 8 \cdot 5 \cdot 3$

5. $12 + 20 + 10 + 5$

6. $26 + 8 + 4 + 22$

7. $3\frac{1}{2} + 4 + 2\frac{1}{2} + 3$

8. $\frac{3}{4} \cdot 12 \cdot 4 \cdot 2$

9. $3.5 + 2.4 + 3.6 + 4.2$

10. $4\frac{1}{2} + 5 + \frac{1}{2} + 3$

11. $0.5 \cdot 2.8 \cdot 4$

12. $2.5 + 2.4 + 2.5 + 3.6$

13. $\frac{4}{5} \cdot 18 \cdot 25 \cdot \frac{2}{9}$

14. $32 \cdot \frac{1}{5} \cdot \frac{1}{2} \cdot 10$

15. $\frac{1}{4} \cdot 7 \cdot 16 \cdot \frac{1}{7}$

16. $3.5 + 8 + 2.5 + 2$

17. $18 \cdot 8 \cdot \frac{1}{2} \cdot \frac{1}{9}$

18. $\frac{3}{4} \cdot 10 \cdot 16 \cdot \frac{1}{2}$

Study Guide and Intervention   Glencoe Algebra 1

1-6 Study Guide and Intervention (continued)

Commutative and Associative Properties

Simplify Expressions The Commutative and Associative Properties can be used along with other properties when evaluating and simplifying expressions.

Example
Simplify $8(y + 2x) + 7y$.

$8(y + 2x) + 7y = 8y + 16x + 7y$ Distributive Property
$ = 8y + 7y + 16x$ Commutative (+)
$ = (8 + 7)y + 16x$ Distributive Property
$ = 15y + 16x$ Substitution

The simplified expression is $15y + 16x$.

Exercises

Simplify each expression.

1. $4x + 3y + x$

2. $3a + 4b + a$

3. $8rs + 2rs^2 + 7rs$

4. $3a^2 + 4b + 10a^2$

5. $6(x + y) + 2(2x + y)$

6. $6n + 2(4n + 5)$

7. $6(a + b) + a + 3b$

8. $5(2x + 3y) + 6(y + x)$

9. $5(0.3x + 0.1y) + 0.2x$

10. $\frac{2}{3} + \frac{1}{2}(x + 10) + \frac{4}{3}$

11. $z^2 + 9x^2 + \frac{4}{3}z^2 + \frac{1}{3}x^2$

12. $6(2x + 4y) + 2(x + 9)$

Write an algebraic expression for each verbal expression. Then simplify.

13. twice the sum of y and z is increased by y

14. four times the product of x and y decreased by $2xy$

15. the product of five and the square of a, increased by the sum of eight, a^2, and 4

16. three times the sum of x and y increased by twice the sum of x and y

NAME _____ DATE _____ PERIOD _____

1-7 Study Guide and Intervention

Logical Reasoning and Counterexamples

Conditional Statements A **conditional statement** is a statement of the form *If A, then B*. Statements in this form are called **if-then statements**. The part of the statement immediately following the word *if* is called the **hypothesis**. The part of the statement immediately following the word *then* is called the **conclusion**.

Example 1 Identify the hypothesis and conclusion of each statement.

a. If it is Wednesday, then Jerri has aerobics class.
 Hypothesis: it is Wednesday
 Conclusion: Jerri has aerobics class

b. If $2x - 4 < 10$, then $x < 7$.
 Hypothesis: $2x - 4 < 10$
 Conclusion: $x < 7$

Example 2 Identify the hypothesis and conclusion of each statement. Then write the statement in if-then form.

a. You and Marylynn can watch a movie on Thursday.
 Hypothesis: it is Thursday
 Conclusion: you and Marylynn can watch a movie
 If it is Thursday, then you and Marylynn can watch a movie.

b. For a number a such that $3a + 2 = 11$, $a = 3$.
 Hypothesis: $3a + 2 = 11$
 Conclusion: $a = 3$
 If $3a + 2 = 11$, then $a = 3$.

Exercises

Identify the hypothesis and conclusion of each statement.

1. If it is April, then it might rain.

2. If you are a sprinter, then you can run fast.

3. If $12 - 4x = 4$, then $x = 2$.

4. If it is Monday, then you are in school.

5. If the area of a square is 49, then the square has side length 7.

Identify the hypothesis and conclusion of each statement. Then write the statement in if-then form.

6. A quadrilateral with equal sides is a rhombus.

7. A number that is divisible by 8 is also divisible by 4.

8. Karlyn goes to the movies when she does not have homework.

1-7 Study Guide and Intervention (continued)

Logical Reasoning and Counterexamples

Deductive Reasoning and Counterexamples Deductive reasoning is the process of using facts, rules, definitions, or properties to reach a valid conclusion. To show that a conditional statement is false, use a **counterexample**, one example for which the conditional statement is false. You need to find only one counterexample for the statement to be false.

Example 1 Determine a valid conclusion from the statement *If two numbers are even, then their sum is even* for the given conditions. If a valid conclusion does not follow, write *no valid conclusion* and explain why.

a. **The two numbers are 4 and 8.**
   4 and 8 are even, and 4 + 8 = 12. Conclusion: The sum of 4 and 8 is even.

b. **The sum of two numbers is 20.**
   Consider 13 and 7. 13 + 7 = 20
   However, 12 + 8, 19 + 1, and 18 + 2 all equal 20. There is no way to determine the two numbers. Therefore there is no valid conclusion.

Example 2 Provide a counterexample to this conditional statement. *If you use a calculator for a math problem, then you will get the answer correct.*
Counterexample: If the problem is 475 ÷ 5 and you press 475 − 5, you will not get the correct answer.

Exercises

Determine a valid conclusion that follows from the statement *If the last digit of a number is 0 or 5, then the number is divisible by 5* for the given conditions. If a valid conclusion does not follow, write *no valid conclusion* and explain why.

1. The number is 120.

2. The number is a multiple of 4.

3. The number is 101.

Find a counterexample for each statement.

4. If Susan is in school, then she is in math class.

5. If a number is a square, then it is divisible by 2.

6. If a quadrilateral has 4 right angles, then the quadrilateral is a square.

7. If you were born in New York, then you live in New York.

8. If three times a number is greater than 15, then the number must be greater than six.

9. If $3x - 2 \leq 10$, then $x < 4$.

NAME _____ DATE _____ PERIOD _____

1-8 Study Guide and Intervention

Number Systems

Square Roots A **square root** is one of two equal factors of a number. For example, the square roots of 36 are 6 and -6, since $6 \cdot 6$ or 6^2 is 36 and $(-6)(-6)$ or $(-6)^2$ is also 36. A rational number like 36, whose square root is a rational number, is called a **perfect square**.

The symbol $\sqrt{}$ is a **radical sign**. It indicates the nonnegative, or **principal**, square root of the number under the radical sign. So $\sqrt{36} = 6$ and $-\sqrt{36} = -6$. The symbol $\pm\sqrt{36}$ represents both square roots.

Example 1 Find $-\sqrt{\dfrac{25}{49}}$.

$-\sqrt{\dfrac{25}{49}}$ represents the negative square root of $\dfrac{25}{49}$.

$\dfrac{25}{49} = \left(\dfrac{5}{7}\right)^2 \rightarrow -\sqrt{\dfrac{25}{49}} = -\dfrac{5}{7}$

Example 2 Find $\pm\sqrt{0.16}$.

$\pm\sqrt{0.16}$ represents the positive and negative square roots of 0.16.
$0.16 = 0.4^2$ and $0.16 = (-0.4)^2$
$\pm\sqrt{0.16} = \pm 0.4$

Exercises

Find each square root.

1. $\sqrt{64}$

2. $-\sqrt{81}$

3. $\sqrt{16.81}$

4. $\pm\sqrt{100}$

5. $-\sqrt{\dfrac{4}{25}}$

6. $-\sqrt{121}$

7. $\pm\sqrt{\dfrac{25}{144}}$

8. $-\sqrt{\dfrac{25}{16}}$

9. $\pm\sqrt{\dfrac{121}{100}}$

10. $-\sqrt{3600}$

11. $-\sqrt{6.25}$

12. $\pm\sqrt{0.0004}$

13. $\sqrt{\dfrac{144}{196}}$

14. $-\sqrt{\dfrac{36}{49}}$

15. $\pm\sqrt{1.21}$

NAME _____ DATE _____ PERIOD _____

1-8 Study Guide and Intervention (continued)

Number Systems

Classify and Order Numbers Numbers such as $\sqrt{2}$ and $\sqrt{3}$ are not perfect squares. Notice what happens when you find these square roots with your calculator. The numbers continue indefinitely without any pattern of repeating digits. Numbers that cannot be written as a terminating or repeating decimal are called **irrational numbers**. The set of **real numbers** consists of the set of irrational numbers and the set of rational numbers together. The chart below illustrates the various kinds of real numbers.

Natural Numbers	{1, 2, 3, 4, ...}
Whole Numbers	{0, 1, 2, 3, 4, ...}
Integers	{..., −3, −2, −1, 0, 1, 2, 3, ...}
Rational Numbers	{all numbers that can be expressed in the form $\frac{a}{b}$, where a and b are integers and $b \neq 0$}
Irrational Numbers	{all numbers that cannot be expressed in the form $\frac{a}{b}$, where a and b are integers and $b \neq 0$}

Example
Name the set or sets of numbers to which each real number belongs.

a. $\frac{4}{11}$ Because 4 and 11 are integers, this number is a rational number.

b. $\sqrt{81}$ Because $\sqrt{81} = 9$, this number is a natural number, a whole number, an integer, and a rational number.

c. $\sqrt{32}$ Because $\sqrt{32} = 5.656854249...$, which is not a repeating or terminating decimal, this number is irrational.

Exercises

Name the set or sets of numbers to which each real number belongs.

1. $\frac{84}{12}$ 　　　 2. $-\frac{6}{7}$ 　　　 3. $\frac{2}{3}$ 　　　 4. $\sqrt{54}$

5. 3.145 　　　 6. $\sqrt{25}$ 　　　 7. 0.62626262... 　　　 8. $\sqrt{22.51}$

Write each set of numbers in order from least to greatest.

9. $-\frac{3}{4}, -5, \sqrt{25}, \frac{7}{4}$ 　　　 10. $\sqrt{0.09}, -0.3131..., \frac{3}{5}$ 　　　 11. $-1.\overline{25}, 0.05, -\frac{1}{4}, \sqrt{5}$

12. $\frac{5}{4}, -2, \sqrt{124}, -3.11$ 　　　 13. $-\sqrt{1.44}, -0.35\frac{1}{5}$ 　　　 14. $0.\overline{35}, 2\frac{1}{3}, -\frac{9}{5}, \sqrt{5}$

NAME _____ DATE _____ PERIOD _____

1-9 Study Guide and Intervention

Functions and Graphs

Interpret Graphs A **function** is a relationship between input and output values. In a function, there is exactly one output for each input. The input values are associated with the **independent variable**, and the output values are associated with the **dependent variable**. Functions can be graphed without using a scale to show the general shape of the graph that represents the function.

Example 1 The graph below represents the height of a football after it is kicked downfield. Identify the independent and the dependent variable. Then describe what is happening in the graph.

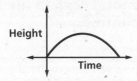

The independent variable is time, and the dependent variable is height. The football starts on the ground when it is kicked. It gains altitude until it reaches a maximum height, then it loses altitude until it falls to the ground.

Example 2 The graph below represents the price of stock over time. Identify the independent and dependent variable. Then describe what is happening in the graph.

The independent variable is time and the dependent variable is price. The price increases steadily, then it falls, then increases, then falls again.

Exercises

1. The graph represents the speed of a car as it travels to the grocery store. Identify the independent and dependent variable. Then describe what is happening in the graph.

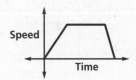

2. The graph represents the balance of a savings account over time. Identify the independent and the dependent variable. Then describe what is happening in the graph.

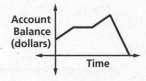

3. The graph represents the height of a baseball after it is hit. Identify the independent and the dependent variable. Then describe what is happening in the graph.

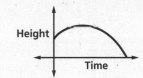

Study Guide and Intervention   17   Glencoe Algebra 1

NAME _____ DATE _____ PERIOD _____

1-9 Study Guide and Intervention (continued)

Functions and Graphs

Draw Graphs You can represent the graph of a function using a **coordinate system**. Input and output values are represented on the graph using **ordered pairs** of the form (x, y). The x-value, called the **x-coordinate**, corresponds to the x-axis, and the y-value, or **y-coordinate** corresponds to the y-axis. A **discrete function** is a function whose graph consists of points that are not connected. When a function can be graphed with a line or smooth curve, it is a **continuous function**.

Example A music store advertises that if you buy 3 CDs at the regular price of $16, then you will receive one CD of the same or lesser value free.

a. Make a table showing the cost of buying 1 to 5 CDs.

Number of CDs	1	2	3	4	5
Total Cost ($)	16	32	48	48	64

b. Write the data as a set of ordered pairs.
(1, 16), (2, 32), (3, 48), (4, 48), (5, 64)

c. Draw a graph that shows the relationship between the number of CDs and the total cost. Is the function discrete or continuous?

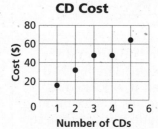

The function is discrete.

Exercises

1. The table below represents the length of a baby versus its age in months.

Age (months)	0	1	2	3	4
Length (inches)	20	21	23	23	24

a. Identify the independent and dependent variables.

b. Write a set of ordered pairs representing the data in the table.

c. Draw a graph showing the relationship between age and length.

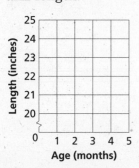

2. The table below represents the value of a car versus its age.

Age (years)	0	1	2	3	4
Value ($)	20,000	18,000	16,000	14,000	13,000

a. Identify the independent and dependent variables.

b. Draw a graph showing the relationship between age and value. Is the function discrete or continuous?

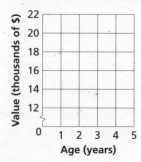

Study Guide and Intervention   Glencoe Algebra 1

2-1 Study Guide and Intervention

Writing Equations

Write Equations Writing equations is one strategy for solving problems. You can use a variable to represent an unspecified number or measure referred to in a problem. Then you can write a verbal expression as an algebraic expression.

Example 1 Translate each sentence into an equation or a formula.

a. Ten times a number x is equal to 2.8 times the difference y minus z.
$10 \times x = 2.8 \times (y - z)$
The equation is $10x = 2.8(y - z)$.

b. A number m minus 8 is the same as a number n divided by 2.
$m - 8 = n \div 2$
The equation is $m - 8 = \frac{n}{2}$.

c. The area of a rectangle equals the length times the width. Translate this sentence into a formula.
Let A = area, ℓ = length, and w = width.
Formula: *Area equals length times width.*
$A = \ell \times w$
The formula for the area of a rectangle is $A = \ell w$.

Example 2 Use the Four-Step Problem-Solving Plan.
The population of the United States in 2005 was about 297,000,000, and the land area of the United States is about 3,500,000 square miles. Find the average number of people per square mile in the United States.
Source: www.census.gov

Step 1 *Explore* You know that there are 297,000,000 people. You want to know the number of people per square mile.

Step 2 *Plan* Write an equation to represent the situation. Let p represent the number of people per square mile.
$3,500,000 \times p = 297,000,000$

Step 3 *Solve* $3,500,000 \times p = 297,000,000$.
$3,500,000p = 297,000,000$ Divide each side by 3,500,000.
$p \approx 84.86$
There about 85 people per square mile.

Step 4 *Check* If there are 85 people per square mile and there are 3,500,000 square miles, $85 \times 3,500,000 = 297,500,000$, or about 297,000,000 people. The answer makes sense.

Exercises

Translate each sentence into an equation or formula.

1. Three times a number t minus twelve equals forty.

2. One-half of the difference of a and b is 54.

3. Three times the sum of d and 4 is 32.

4. The area A of a circle is the product of π and the radius r squared.

WEIGHT LOSS For Exercises 5–6, use the following information.

Lou wants to lose weight to audition for a part in a play. He weighs 160 pounds now. He wants to weigh 150 pounds.

5. If p represents the number of pounds he wants to lose, write an equation to represent this situation.

6. How many pounds does he need to lose to reach his goal?

2-1 Study Guide and Intervention (continued)

Writing Equations

Write Verbal Sentences You can translate equations into verbal sentences.

Example Translate each equation into a verbal sentence.

a. $4n - 8 = 12$.

$$\underbrace{4n}\ \underbrace{-}\ \underbrace{8}\ \underbrace{=}\ \underbrace{12}$$
Four times n minus eight equals twelve.

b. $a^2 + b^2 = c^2$

$$\underbrace{a^2 + b^2}\ \underbrace{=}\ \underbrace{c^2}$$
The sum of the squares of a and b is equal to the square of c.

Exercises

Translate each equation into a verbal sentence.

1. $4a - 5 = 23$

2. $10 + k = 4k$

3. $6xy = 24$

4. $x^2 + y^2 = 8$

5. $p + 3 = 2p$

6. $b = \frac{1}{3}(h - 1)$

7. $100 - 2x = 80$

8. $3(g + h) = 12$

9. $p^2 - 2p = 9$

10. $C = \frac{5}{9}(F - 32)$

11. $V = \frac{1}{3}Bh$

12. $A = \frac{1}{2}hb$

NAME _____ DATE _____ PERIOD _____

2-2 Study Guide and Intervention

Solving Equations by Using Addition and Subtraction

Solve Using Addition If the same number is added to each side of an equation, the resulting equation is equivalent to the original one. In general if the original equation involves subtraction, this property will help you solve the equation.

| Addition Property of Equality | For any numbers a, b, and c, if $a = b$, then $a + c = b + c$. |

Example 1 Solve $m - 32 = 18$.

$m - 32 = 18$ Original equation
$m - 32 + 32 = 18 + 32$ Add 32 to each side.
$m = 50$ Simplify.

The solution is 50.

Example 2 Solve $-18 = p - 12$.

$-18 = p - 12$ Original equation
$-18 + 12 = p - 12 + 12$ Add 12 to each side.
$p = -6$ Simplify.

The solution is -6.

Exercises

Solve each equation. Then check your solution.

1. $h - 3 = -2$
2. $m - 8 = -12$
3. $p - 5 = 15$

4. $20 = y - 8$
5. $k - 0.5 = 2.3$
6. $w - \frac{1}{2} = \frac{5}{8}$

7. $h - 18 = -17$
8. $-12 = -24 + k$
9. $j - 0.2 = 1.8$

10. $b - 40 = -40$
11. $m - (-12) = 10$
12. $w - \frac{3}{2} = \frac{1}{4}$

Write an equation for each problem. Then solve the equation and check the solution.

13. Twelve subtracted from a number equals 25. Find the number.

14. What number decreased by 52 equals -12?

15. Fifty subtracted from a number equals eighty. Find the number.

16. What number minus one-half is equal to negative one-half?

17. The difference of a number and eight is equal to 14. What is the number?

18. A number decreased by fourteen is equal to eighteen. What is the number?

NAME _____ DATE _____ PERIOD _____

2-2 Study Guide and Intervention (continued)
Solving Equations by Using Addition and Subtraction

Solve Using Subtraction If the same number is subtracted from each side of an equation, the resulting equation is equivalent to the original one. In general if the original equation involves addition, this property will help you solve the equation.

Subtraction Property of Equality	For any numbers a, b, and c, if $a = b$, then $a - c = b - c$.

Example Solve $22 + p = -12$.

$22 + p = -12$ Original equation
$22 + p - 22 = -12 - 22$ Subtract 22 from each side.
$p = -34$ Simplify.

The solution is -34.

Exercises

Solve each equation. Then check your solution.

1. $x + 12 = 6$

2. $z + 2 = -13$

3. $-17 = b + 4$

4. $s + (-9) = 7$

5. $-3.2 = \ell + (-0.2)$

6. $-\dfrac{3}{8} + x = \dfrac{5}{8}$

7. $19 + h = -4$

8. $-12 = k + 24$

9. $j + 1.2 = 2.8$

10. $b + 80 = -80$

11. $m + (-8) = 2$

12. $w + \dfrac{3}{2} = \dfrac{5}{8}$

Write an equation for each problem. Then solve the equation and check the solution.

13. Twelve added to a number equals 18. Find the number.

14. What number increased by 20 equals -10?

15. The sum of a number and fifty equals eighty. Find the number.

16. What number plus one-half is equal to four?

17. The sum of a number and 3 is equal to -15. What is the number?

2-3 Study Guide and Intervention

Solving Equations by using Multiplication and Division

Solve Using Multiplication If each side of an equation is multiplied by the same number, the resulting equation is equivalent to the given one. You can use the property to solve equations involving multiplication and division.

Multiplication Property of Equality	For any numbers a, b, and c, if $a = b$, then $ac = bc$.

Example 1 Solve $3\frac{1}{2}p = 1\frac{1}{2}$.

$3\frac{1}{2}p = 1\frac{1}{2}$ Original equation

$\frac{7}{2}p = \frac{3}{2}$ Rewrite each mixed number as an improper fraction.

$\frac{2}{7}\left(\frac{7}{2}p\right) = \frac{2}{7}\left(\frac{3}{2}\right)$ Multiply each side by $\frac{2}{7}$.

$p = \frac{3}{7}$ Simplify.

The solution is $\frac{3}{7}$.

Example 2 Solve $-\frac{1}{4}n = 16$.

$-\frac{1}{4}n = 16$ Original equation

$-4\left(-\frac{1}{4}n\right) = -4(16)$ Multiply each side by -4.

$n = -64$ Simplify.

The solution is -64.

Exercises

Solve each equation. Then check your solution.

1. $\frac{h}{3} = -2$

2. $\frac{1}{8}m = 6$

3. $\frac{1}{5}p = \frac{3}{5}$

4. $5 = \frac{y}{12}$

5. $-\frac{1}{4}k = -2.5$

6. $-\frac{m}{8} = \frac{5}{8}$

7. $-1\frac{1}{2}h = 4$

8. $-12 = -\frac{3}{2}k$

9. $\frac{j}{3} = \frac{2}{5}$

10. $-3\frac{1}{3}b = 5$

11. $\frac{7}{10}m = 10$

12. $\frac{p}{5} = -\frac{1}{4}$

Write an equation for each problem. Then solve the equation.

13. One-fifth of a number equals 25. Find the number.

14. What number divided by 2 equals -18?

15. A number divided by eight equals 3. Find the number.

16. One and a half times a number equals 6. Find the number.

NAME _____ DATE _____ PERIOD _____

2-3 Study Guide and Intervention (continued)

Solving Equations by Using Multiplication and Division

Solve Using Division To solve equations with multiplication and division, you can also use the Division Property of Equality. If each side of an equation is divided by the same number, the resulting equation is true.

Division Property of Equality	For any numbers a, b, and c, with $c \neq 0$, if $a = b$, then $\frac{a}{c} = \frac{b}{c}$.

Example 1 Solve $8n = 64$.

$8n = 64$ Original equation
$\frac{8n}{8} = \frac{64}{8}$ Divide each side by 8.
$n = 8$ Simplify.

The solution is 8.

Example 2 Solve $-5n = 60$.

$-5n = 60$ Original equation
$\frac{-5n}{-5} = \frac{60}{-5}$ Divide each side by -5.
$n = -12$ Simplify.

The solution is -12.

Exercises

Solve each equation. Then check your solution.

1. $3h = -42$
2. $8m = 16$
3. $-3t = 51$

4. $-3r = -24$
5. $8k = -64$
6. $-2m = 16$

7. $12h = 4$
8. $-2.4p = 7.2$
9. $0.5j = 5$

10. $-25 = 5m$
11. $6m = 15$
12. $-1.5p = -75$

Write an equation for each problem. Then solve the equation.

13. Four times a number equals 64. Find the number.

14. What number multiplied by -4 equals -16?

15. A number times eight equals -36. Find the number.

NAME _____ DATE _____ PERIOD _____

2-4 Study Guide and Intervention

Solving Multi-Step Equations

Work Backward Working backward is one of many problem-solving strategies that you can use to solve problems. To work backward, start with the result given at the end of a problem and undo each step to arrive at the beginning number.

Example 1 A number is divided by 2, and then 8 is subtracted from the quotient. The result is 16. What is the number?

Solve the problem by working backward.
The final number is 16. Undo subtracting 8 by adding 8 to get 24. To undo dividing 24 by 2, multiply 24 by 2 to get 48.
The original number is 48.

Example 2 A bacteria culture doubles each half hour. After 3 hours, there are 6400 bacteria. How many bacteria were there to begin with?

Solve the problem by working backward.
The bacteria have grown for 3 hours. Since there are 2 one-half hour periods in one hour, in 3 hours there are 6 one-half hour periods. Since the bacteria culture has grown for 6 time periods, it has doubled 6 times. Undo the doubling by halving the number of bacteria 6 times.

$$6{,}400 \times \frac{1}{2} \times \frac{1}{2} \times \frac{1}{2} \times \frac{1}{2} \times \frac{1}{2} \times \frac{1}{2} = 6{,}400 \times \frac{1}{64}$$
$$= 100$$

There were 100 bacteria to begin with.

Exercises

Solve each problem by working backward.

1. A number is divided by 3, and then 4 is added to the quotient. The result is 8. Find the number.

2. A number is multiplied by 5, and then 3 is subtracted from the product. The result is 12. Find the number.

3. Eight is subtracted from a number, and then the difference is multiplied by 2. The result is 24. Find the number.

4. Three times a number plus 3 is 24. Find the number.

5. **CAR RENTAL** Angela rented a car for $29.99 a day plus a one-time insurance cost of $5.00. Her bill was $124.96. For how many days did she rent the car?

6. **MONEY** Mike withdrew an amount of money from his bank account. He spent one fourth for gasoline and had $90 left. How much money did he withdraw?

7. **TELEVISIONS** In 2003, 68% of households with TVs subscribed to cable TV. If 8,000 more subscribers were added to the number of households with cable, the total number of households with cable TV would be 73,448,000. How many households were there with TV in 2003? **Source:** World Almanac

Study Guide and Intervention · 25 · Glencoe Algebra 1

NAME _____ DATE _____ PERIOD _____

2-4 Study Guide and Intervention (continued)

Solving Multi-Step Equations

Solve Multi-Step Equations To solve equations with more than one operation, often called **multi-step equations**, undo operations by working backward. Reverse the usual order of operations as you work.

Example
Solve $5x + 3 = 23$.

$5x + 3 = 23$	Original equation.
$5x + 3 - 3 = 23 - 3$	Subtract 3 from each side.
$5x = 20$	Simplify.
$\dfrac{5x}{5} = \dfrac{20}{5}$	Divide each side by 5.
$x = 4$	Simplify.

Exercises

Solve each equation. Then check your solution.

1. $5x + 2 = 27$
2. $6x + 9 = 27$
3. $5x + 16 = 51$

4. $14n - 8 = 34$
5. $0.6x - 1.5 = 1.8$
6. $\dfrac{7}{8}p - 4 = 10$

7. $16 = \dfrac{d - 12}{14}$
8. $8 + \dfrac{3n}{12} = 13$
9. $\dfrac{g}{-5} + 3 = -13$

10. $\dfrac{4b + 8}{-2} = 10$
11. $0.2x - 8 = -2$
12. $3.2y - 1.8 = 3$

13. $-4 = \dfrac{7x - (-1)}{-8}$
14. $8 = -12 + \dfrac{k}{-4}$
15. $0 = 10y - 40$

Write an equation and solve each problem.

16. Find three consecutive integers whose sum is 96.

17. Find two consecutive odd integers whose sum is 176.

18. Find three consecutive integers whose sum is -93.

2-5 Study Guide and Intervention

Solving Equations with the Variable on Each Side

Variables on Each Side To solve an equation with the same variable on each side, first use the Addition or the Subtraction Property of Equality to write an equivalent equation that has the variable on just one side of the equation. Then solve the equation.

Example 1 Solve $5y - 8 = 3y + 12$.

$$5y - 8 = 3y + 12$$
$$5y - 8 - 3y = 3y + 12 - 3y$$
$$2y - 8 = 12$$
$$2y - 8 + 8 = 12 + 8$$
$$2y = 20$$
$$\frac{2y}{2} = \frac{20}{2}$$
$$y = 10$$

The solution is 10.

Example 2 Solve $-11 - 3y = 8y + 1$.

$$-11 - 3y = 8y + 1$$
$$-11 - 3y + 3y = 8y + 1 + 3y$$
$$-11 = 11y + 1$$
$$-11 - 1 = 11y + 1 - 1$$
$$-12 = 11y$$
$$\frac{-12}{11} = \frac{11y}{11}$$
$$-1\frac{1}{11} = y$$

The solution is $-1\frac{1}{11}$.

Exercises

Solve each equation. Then check your solution.

1. $6 - b = 5b + 30$

2. $5y - 2y = 3y + 2$

3. $5x + 2 = 2x - 10$

4. $4n - 8 = 3n + 2$

5. $1.2x + 4.3 = 2.1 - x$

6. $4.4s + 6.2 = 8.8s - 1.8$

7. $\frac{1}{2}b + 4 = \frac{1}{8}b + 88$

8. $\frac{3}{4}k - 5 = \frac{1}{4}k - 1$

9. $8 - 5p = 4p - 1$

10. $4b - 8 = 10 - 2b$

11. $0.2x - 8 = -2 - x$

12. $3y - 1.8 = 3y - 1.8$

13. $-4 - 3x = 7x - 6$

14. $8 + 4k = -10 + k$

15. $20 - a = 10a - 2$

16. $\frac{2}{3}n + 8 = \frac{1}{2}n + 2$

17. $\frac{2}{5}y - 8 = 9 - \frac{3}{5}y$

18. $-4r + 5 = 5 - 4r$

19. $-4 - 3x = 6x - 6$

20. $18 - 4k = -10 - 4k$

21. $12 + 2y = 10y - 12$

NAME _____ DATE _____ PERIOD _____

2-5 Study Guide and Intervention (continued)

Solving Equations with the Variable on Each Side

Grouping Symbols When solving equations that contain grouping symbols, first use the Distributive Property to eliminate grouping symbols. Then solve.

Example Solve $4(2a - 1) = -10(a - 5)$.

$4(2a - 1) = -10(a - 5)$	Original equation
$8a - 4 = -10a + 50$	Distributive Property
$8a - 4 + 10a = -10a + 50 + 10a$	Add 10a to each side.
$18a - 4 = 50$	Simplify.
$18a - 4 + 4 = 50 + 4$	Add 4 to each side.
$18a = 54$	Simplify.
$\dfrac{18a}{18} = \dfrac{54}{18}$	Divide each side by 18.
$a = 3$	Simplify.

The solution is 3.

Exercises

Solve each equation. Then check your solution.

1. $-3(x + 5) = 3(x - 1)$

2. $2(7 + 3t) = -t$

3. $3(a + 1) - 5 = 3a - 2$

4. $75 - 9g = 5(-4 + 2g)$

5. $5(f + 2) = 2(3 - f)$

6. $4(p + 3) = 36$

7. $18 = 3(2c + 2)$

8. $3(d - 8) = 3d$

9. $5(p + 3) + 9 = 3(p - 2) + 6$

10. $4(b - 2) = 2(5 - b)$

11. $1.2(x - 2) = 2 - x$

12. $\dfrac{3 + y}{4} = \dfrac{-y}{8}$

13. $\dfrac{a - 8}{12} = \dfrac{2a + 5}{3}$

14. $2(4 + 2k) + 10 = k$

15. $2(w - 1) + 4 = 4(w + 1)$

16. $6(n - 1) = 2(2n + 4)$

17. $2[2 + 3(y - 1)] = 22$

18. $-4(r + 2) = 4(2 - 4r)$

19. $-3(x - 8) = 24$

20. $4(4 - 4k) = -10 - 16k$

21. $6(2 - 2y) = 5(2y - 2)$

NAME _____ DATE _____ PERIOD _____

2-6 Study Guide and Intervention

Ratios and Proportions

Ratios and Proportions A **ratio** is a comparison of two numbers by division. The ratio of x to y can be expressed as x to y, $x{:}y$ or $\dfrac{x}{y}$. Ratios are usually expressed in simplest form.

An equation stating that two ratios are equal is called a **proportion**. To determine whether two ratios form a proportion, express both ratios in simplest form or check cross products.

Example 1 Determine whether the ratios $\dfrac{24}{36}$ and $\dfrac{12}{18}$ form a proportion.

$\dfrac{24}{36} = \dfrac{2}{3}$ when expressed in simplest form.

$\dfrac{12}{18} = \dfrac{2}{3}$ when expressed in simplest form.

The ratios $\dfrac{24}{36}$ and $\dfrac{12}{18}$ form a proportion because they are equal when expressed in simplest form.

Example 2 Use cross products to determine whether $\dfrac{10}{18}$ and $\dfrac{25}{45}$ form a proportion.

$\dfrac{10}{18} \stackrel{?}{=} \dfrac{25}{45}$ Write the proportion.

$10(45) \stackrel{?}{=} 18(25)$ Cross products

$450 = 450$ Simplify.

The cross products are equal, so $\dfrac{10}{18} = \dfrac{25}{45}$. Since the ratios are equal, they form a proportion.

Exercises

Use cross products to determine whether each pair of ratios forms a proportion.

1. $\dfrac{1}{2}, \dfrac{16}{32}$

2. $\dfrac{5}{8}, \dfrac{10}{15}$

3. $\dfrac{10}{20}, \dfrac{25}{49}$

4. $\dfrac{25}{36}, \dfrac{15}{20}$

5. $\dfrac{12}{32}, \dfrac{3}{16}$

6. $\dfrac{4}{9}, \dfrac{12}{27}$

7. $\dfrac{0.1}{2}, \dfrac{5}{100}$

8. $\dfrac{15}{20}, \dfrac{9}{12}$

9. $\dfrac{14}{21}, \dfrac{20}{30}$

10. $2{:}3$, $20{:}30$

11. 5 to 9, 25 to 45

12. $\dfrac{72}{64}, \dfrac{9}{8}$

13. $5{:}5$, $30{:}20$

14. 18 to 24, 50 to 75

15. $100{:}75$, $44{:}33$

16. $\dfrac{0.05}{1}, \dfrac{1}{20}$

17. $\dfrac{1.5}{2}, \dfrac{6}{8}$

18. $\dfrac{0.1}{0.2}, \dfrac{0.45}{0.9}$

2-6 Study Guide and Intervention (continued)
Ratios and Proportions

Solve Proportions If a proportion involves a variable, you can use cross products to solve the proportion. In the proportion $\frac{x}{5} = \frac{10}{13}$, x and 13 are called **extremes**. They are the first and last terms of the proportion. 5 and 10 are called **means**. They are the middle terms of the proportion. In a proportion, the product of the extremes is equal to the product of the means.

Means-Extremes Property of Proportions	For any numbers a, b, c, and d, if $\frac{a}{b} = \frac{c}{d}$, then $ad = bc$.

Example Solve $\frac{x}{5} = \frac{10}{13}$.

$\frac{x}{5} = \frac{10}{13}$ Original proportion

$13(x) = 5(10)$ Cross products

$13x = 50$ Simplify.

$\frac{13x}{13} = \frac{50}{13}$ Divide each side by 13.

$x = 3\frac{11}{13}$ Simplify.

Exercises

Solve each proportion.

1. $\frac{-3}{x} = \frac{2}{8}$

2. $\frac{1}{t} = \frac{5}{3}$

3. $\frac{0.1}{2} = \frac{0.5}{x}$

4. $\frac{x+1}{4} = \frac{3}{4}$

5. $\frac{4}{6} = \frac{8}{x}$

6. $\frac{x}{21} = \frac{3}{63}$

7. $\frac{9}{y+1} = \frac{18}{54}$

8. $\frac{3}{d} = \frac{18}{3}$

9. $\frac{5}{8} = \frac{p}{24}$

10. $\frac{4}{b-2} = \frac{4}{12}$

11. $\frac{1.5}{x} = \frac{12}{x}$

12. $\frac{3+y}{4} = \frac{-y}{8}$

13. $\frac{a-8}{12} = \frac{15}{3}$

14. $\frac{12}{k} = \frac{24}{k}$

15. $\frac{2+w}{6} = \frac{12}{9}$

Use a proportion to solve each problem.

16. **MODELS** To make a model of the Guadeloupe River bed, Hermie used 1 inch of clay for 5 miles of the river's actual length. His model river was 50 inches long. How long is the Guadeloupe River?

17. **EDUCATION** Josh finished 24 math problems in one hour. At that rate, how many hours will it take him to complete 72 problems?

NAME _____ DATE _____ PERIOD _____

2-7 Study Guide and Intervention

Percent of Change

Percent of Change When an increase or decrease in an amount is expressed as a percent, the percent is called the **percent of change**. If the new number is greater than the original number, the percent of change is a **percent of increase**. If the new number is less than the original number, the percent of change is the **percent of decrease**.

Example 1
Find the percent of increase.
 original: 48
 new: 60

First, subtract to find the amount of increase. The amount of increase is $60 - 48 = 12$.

Then find the percent of increase by using the original number, 48, as the base.

$\frac{12}{48} = \frac{r}{100}$ Percent proportion
$12(100) = 48(r)$ Cross products
$1200 = 48r$ Simplify.
$\frac{1200}{48} = \frac{48r}{48}$ Divide each side by 48.
$25 = r$ Simplify.

The percent of increase is 25%.

Example 2
Find the percent of decrease.
 original: 30
 new: 22

First, subtract to find the amount of decrease. The amount of decrease is $30 - 22 = 8$.

Then find the percent of decrease by using the original number, 30, as the base.

$\frac{8}{30} = \frac{r}{100}$ Percent proportion
$8(100) = 30(r)$ Cross products
$800 = 30r$ Simplify.
$\frac{800}{30} = \frac{30r}{30}$ Divide each side by 30.
$26\frac{2}{3} = r$ Simplify.

The percent of decrease is $26\frac{2}{3}$%, or about 27%.

Exercises

State whether each percent of change is a percent of increase or a percent of decrease. Then find each percent of change. Round to the nearest whole percent.

1. original: 50
   new: 80

2. original: 90
   new: 100

3. original: 45
   new: 20

4. original: 77.5
   new: 62

5. original: 140
   new: 150

6. original: 135
   new: 90

7. original: 120
   new: 180

8. original: 90
   new: 270

9. original: 27.5
   new: 25

10. original: 84
    new: 98

11. original: 12.5
    new: 10

12. original: 250
    new: 500

Study Guide and Intervention   Glencoe Algebra 1

NAME _____ DATE _____ PERIOD _____

2-7 Study Guide and Intervention (continued)

Percent of Change

Solve Problems Discounted prices and prices including tax are applications of percent of change. Discount is the amount by which the regular price of an item is reduced. Thus, the discounted price is an example of percent of decrease. Sales tax is amount that is added to the cost of an item, so the price including tax is an example of percent of increase.

Example A coat is on sale for 25% off the original price. If the original price of the coat is $75, what is the discounted price?

The discount is 25% of the original price.

25% of $75 = 0.25 × 75   25% = 0.25
             = 18.75   Use a calculator.

Subtract $18.75 from the original price.
$75 − $18.75 = $56.25

The discounted price of the coat is $56.25.

Exercises

Find the final price of each item. When a discount and a sales tax are listed, compute the discount price before computing the tax.

1. Compact disc: $16
   Discount: 15%

2. Two concert tickets: $28
   Student discount: 28%

3. Airline ticket: $248.00
   Superair discount: 33%

4. Shirt: $24.00
   Sales tax: 4%

5. CD player: $142.00
   Sales tax: 5.5%

6. Celebrity calendar: $10.95
   Sales tax: 7.5%

7. Class ring: $89.00
   Group discount: 17%
   Sales tax: 5%

8. Software: $44.00
   Discount: 21%
   Sales tax: 6%

9. Video recorder: $110.95
   Discount: 20%
   Sales tax: 5%

10. **VIDEOS** The original selling price of a new sports video was $65.00. Due to the demand the price was increased to $87.75. What was the percent of increase over the original price?

11. **SCHOOL** A high school paper increased its sales by 75% when it ran an issue featuring a contest to win a class party. Before the contest issue, 10% of the school's 800 students bought the paper. How many students bought the contest issue?

12. **BASEBALL** Baseball tickets cost $15 for general admission or $20 for box seats. The sales tax on each ticket is 8%, and the municipal tax on each ticket is an additional 10% of the base price. What is the final cost of each type of ticket?

NAME _____ DATE _____ PERIOD _____

2-8 Study Guide and Intervention

Solving Equations and Formulas

Solve for Variables Sometimes you may want to solve an equation such as $V = \ell wh$ for one of its variables. For example, if you know the values of V, w, and h, then the equation $\ell = \dfrac{V}{wh}$ is more useful for finding the value of ℓ. If an equation that contains more than one variable is to be solved for a specific variable, use the properties of equality to isolate the specified variable on one side of the equation.

Example 1 Solve $2x - 4y = 8$ for y.

$2x - 4y = 8$
$2x - 4y - 2x = 8 - 2x$
$-4y = 8 - 2x$
$\dfrac{-4y}{-4} = \dfrac{8 - 2x}{-4}$
$y = \dfrac{8 - 2x}{-4}$ or $\dfrac{2x - 8}{4}$

The value of y is $\dfrac{2x - 8}{4}$.

Example 2 Solve $3m - n = km - 8$ for m.

$3m - n = km - 8$
$3m - n - km = km - 8 - km$
$3m - n - km = -8$
$3m - n - km + n = -8 + n$
$3m - km = -8 + n$
$m(3 - k) = -8 + n$
$\dfrac{m(3 - k)}{3 - k} = \dfrac{-8 + n}{3 - k}$
$m = \dfrac{-8 + n}{3 - k}$, or $\dfrac{n - 8}{3 - k}$

The value of m is $\dfrac{n - 8}{3 - k}$. Since division by 0 is undefined, $3 - k \neq 0$, or $k \neq 3$.

Exercises

Solve each equation or formula for the variable specified.

1. $ax - b = c$ for x

2. $15x + 1 = y$ for x

3. $(x + f) + 2 = j$ for x

4. $xy + z = 9$ for y

5. $x(4 - k) = p$ for k

6. $7x + 3y = m$ for y

7. $4(c + 3) = t$ for c

8. $2x + b = c$ for x

9. $x(1 + y) = z$ for x

10. $16z + 4x = y$ for x

11. $d = rt$ for r

12. $A = \dfrac{h(a + b)}{2}$ for h

13. $C = \dfrac{5}{9}(F - 32)$ for F

14. $P = 2\ell + 2w$ for w

15. $A = \ell w$ for ℓ

NAME _____ DATE _____ PERIOD _____

2-8 Study Guide and Intervention (continued)

Solving Equations and Formulas

Use Formulas Many real-world problems require the use of formulas. Sometimes solving a formula for a specified variable will help solve the problem.

Example The formula $C = \pi d$ represents the circumference of a circle, or the distance around the circle, where d is the diameter. If an airplane could fly around Earth at the equator without stopping, it would have traveled about 24,900 miles. Find the diameter of Earth.

$C = \pi d$ Given formula

$d = \dfrac{C}{\pi}$ Solve for d.

$d = \dfrac{24{,}900}{3.14}$ Use $\pi = 3.14$.

$d \approx 7930$ Simplify.

The diameter of Earth is about 7930 miles.

Exercises

1. **GEOMETRY** The volume of a cylinder V is given by the formula $V = \pi r^2 h$, where r is the radius and h is the height.

   a. Solve the formula for h.

   b. Find the height of a cylinder with volume 2500π feet and radius 10 feet.

2. **WATER PRESSURE** The water pressure on a submerged object is given by $P = 64d$, where P is the pressure in pounds per square foot, and d is the depth of the object in feet.

   a. Solve the formula for d.

   b. Find the depth of a submerged object if the pressure is 672 pounds per square foot.

3. **GRAPHS** The equation of a line containing the points $(a, 0)$ and $(0, b)$ is given by the formula $\dfrac{x}{a} + \dfrac{y}{b} = 1$.

   a. Solve the equation for y.

   b. Suppose the line contains the points $(4, 0)$, and $(0, -2)$. If $x = 3$, find y.

4. **GEOMETRY** The surface area of a rectangular solid is given by the formula $S = 2\ell w + 2\ell h + 2wh$, where ℓ = length, w = width, and h = height.

   a. Solve the formula for h.

   b. The surface area of a rectangular solid with length 6 centimeters and width 3 centimeters is 72 square centimeters. Find the height.

NAME _____ DATE _____ PERIOD _____

2-9 Study Guide and Intervention

Weighted Averages

Mixture Problems

Weighted Average	The weighted average M of a set of data is the sum of the product of each number in the set and its weight divided by the sum of all the weights.

Mixture Problems are problems where two or more parts are combined into a whole. They involve weighted averages. In a mixture problem, the weight is usually a price or a percent of something.

Example Delectable Cookie Company sells chocolate chip cookies for $6.95 per pound and white chocolate cookies for $5.95 per pound. How many pounds of chocolate chip cookies should be mixed with 4 pounds of white chocolate cookies to obtain a mixture that sells for $6.75 per pound.

Let w = the number of pounds of chocolate chip cookies

	Number of Pounds	Price per Pound	Total Price
Chocolate Chip	w	6.95	6.95w
White Chocolate	4	5.95	4(5.95)
Mixture	$w + 4$	6.75	6.75(w + 4)

Equation: $6.95w + 4(5.95) = 6.75(w + 4)$

Solve the equation.

$6.95w + 4(5.95) = 6.75(w + 4)$	Original equation
$6.95w + 23.80 = 6.75w + 27$	Simplify.
$6.95w + 23.80 - 6.75w = 6.75w + 27 - 6.75w$	Subtract 6.75w from each side.
$0.2w + 23.80 = 27$	Simplify.
$0.2w + 23.80 - 23.80 = 27 - 23.80$	Subtract 23.80 from each side.
$0.2w = 3.2$	Simplify.
$w = 16$	Simplify.

16 pounds of chocolate chip cookies should be mixed with 4 pounds of white chocolate cookies.

Exercises

1. **SOLUTIONS** How many grams of sugar must be added to 60 grams of a solution that is 32% sugar to obtain a solution that is 50% sugar?

2. **NUTS** The Quik Mart has two kinds of nuts. Pecans sell for $1.55 per pound and walnuts sell for $1.95 per pound. How many pounds of walnuts must be added to 15 pounds of pecans to make a mixture that sells for $1.75 per pound?

3. **INVESTMENTS** Alice Gleason invested a portion of $32,000 at 9% interest and the balance at 11% interest. How much did she invest at each rate if her total income from both investments was $3,200.

4. **MILK** Whole milk is 4% butterfat. How much skim milk with 0% butterfat should be added to 32 ounces of whole milk to obtain a mixture that is 2.5% butterfat?

2-9 Study Guide and Intervention (continued)

Weighted Averages

Uniform Motion Problems Motion problems are another application of weighted averages. **Uniform motion problems** are problems where an object moves at a certain speed, or rate. Use the formula $d = rt$ to solve these problems, where d is the distance, r is the rate, and t is the time.

Example Bill Gutierrez drove at a speed of 65 miles per hour on an expressway for 2 hours. He then drove for 1.5 hours at a speed of 45 miles per hour on a state highway. What was his average speed?

$M = \dfrac{65 \cdot 2 + 45 \cdot 1.5}{2 + 1.5}$ Definition of weighted average

$ \approx 56.4$ Simplify.

Bill drove at an average speed of about 56.4 miles per hour.

Exercises

1. **TRAVEL** Mr. Anders and Ms. Rich each drove home from a business meeting. Mr. Anders traveled east at 100 kilometers per hour and Ms. Rich traveled west at 80 kilometers per hours. In how many hours were they 100 kilometers apart.

2. **AIRPLANES** An airplane flies 750 miles due west in $1\frac{1}{2}$ hours and 750 miles due south in 2 hours. What is the average speed of the airplane?

3. **TRACK** Sprinter A runs 100 meters in 15 seconds, while sprinter B starts 1.5 seconds later and runs 100 meters in 14 seconds. If each of them runs at a constant rate, who is further in 10 seconds after the start of the race? Explain.

4. **TRAINS** An express train travels 90 kilometers per hour from Smallville to Megatown. A local train takes 2.5 hours longer to travel the same distance at 50 kilometers per hour. How far apart are Smallville and Megatown?

5. **CYCLING** Two cyclists begin traveling in the same direction on the same bike path. One travels at 15 miles per hour, and the other travels at 12 miles per hour. When will the cyclists be 10 miles apart?

6. **TRAINS** Two trains leave Chicago, one traveling east at 30 miles per hour and one traveling west at 40 miles per hour. When will the trains be 210 miles apart?

NAME _____ DATE _____ PERIOD _____

3-1 Study Guide and Intervention

Representing Relations

Represent Relations A **relation** is a set of ordered pairs. A relation can be represented by a set of ordered pairs, a table, a graph, or a **mapping**. A mapping illustrates how each element of the domain is paired with an element in the range.

Example 1 Express the relation {(1, 1), (0, 2), (3, −2)} as a table, a graph, and a mapping. State the domain and range of the relation.

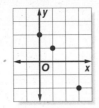

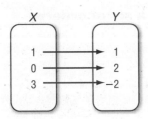

The domain for this relation is {0, 1, 3}. The range for this relation is {−2, 1, 2}.

Example 2 A person playing racquetball uses 4 calories per hour for every pound he or she weighs.

a. Make a table to show the relation between weight and calories burned in one hour for people weighing 100, 110, 120, and 130 pounds.
Source: *The Math Teacher's Book of Lists*

x	y
100	400
110	440
120	480
130	520

b. Give the domain and range.
domain: {100, 110, 120, 130}
range: {400, 440, 480, 520}

c. Graph the relation.

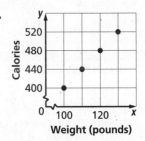

Exercises

1. Express the relation {(−2, −1), (3, 3), (4, 3)} as a table, a graph, and a mapping. Then determine the domain and range.

  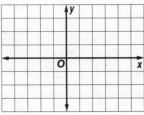

2. The temperature in a house drops 2° for every hour the air conditioner is on between the hours of 6 A.M. and 11 A.M. Make a graph to show the relationship between time and temperature if the temperature at 6 A.M. was 82°F.

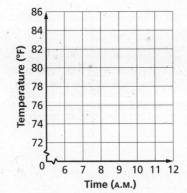

3-1 Study Guide and Intervention (continued)

Representing Relations

Inverse Relations The **inverse** of any relation is obtained by switching the coordinates in each ordered pair.

Example
Express the relation shown in the mapping as a set of ordered pairs. Then write the inverse of the relation.

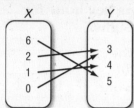

Relation: {(6, 5), (2, 3), (1, 4), (0, 3)}
Inverse: {(5, 6), (3, 2), (4, 1), (3, 0)}

Exercises

Express the relation shown in each table, mapping, or graph as a set of ordered pairs. Then write the inverse of each relation.

1.

x	y
−2	4
−1	3
2	1
4	5

2.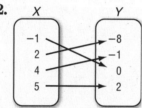

3.

x	y
−3	5
−2	−1
1	0
2	4

4.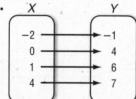

5.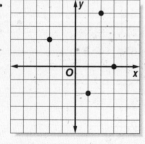

6.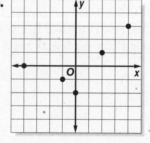

NAME _____ DATE _____ PERIOD _____

3-2 Study Guide and Intervention

Representing Functions

Identify Functions Relations in which each element of the domain is paired with exactly one element of the range are called **functions**.

Example 1 Determine whether the relation {(6, −3), (4, 1), (7, −2), (−3, 1)} is a function. Explain.

Since each element of the domain is paired with exactly one element of the range, this relation is a function.

Example 2 Determine whether $3x - y = 6$ is a function.

Since the equation is in the form $Ax + By = C$, the graph of the equation will be a line, as shown at the right.

If you draw a vertical line through each value of x, the vertical line passes through just one point of the graph. Thus, the line represents a function.

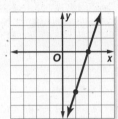

Exercises

Determine whether each relation is a function.

1.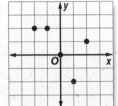

2.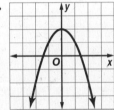

3.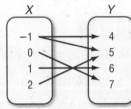

4.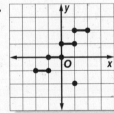

5.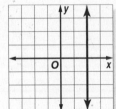

6.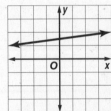

7. {(4, 2), (2, 3), (6, 1)}

8. {(−3, −3), (−3, 4), (−2, 4)}

9. {(−1, 0), (1, 0)}

10. $-2x + 4y = 0$

11. $x^2 + y^2 = 8$

12. $x = -4$

Study Guide and Intervention   39   Glencoe Algebra 1

3-2 Study Guide and Intervention (continued)

Representing Functions

Function Values Equations that are functions can be written in a form called **function notation**. For example, $y = 2x - 1$ can be written as $f(x) = 2x - 1$. In the function, x represents the elements of the domain, and $f(x)$ represents the elements of the range. Suppose you want to find the value in the range that corresponds to the element 2 in the domain. This is written $f(2)$ and is read "f of 2." The value of $f(2)$ is found by substituting 2 for x in the equation.

Example
If $f(x) = 3x - 4$, find each value.

a. $f(3)$

$f(3) = 3(3) - 4$ Replace x with 3.
$= 9 - 4$ Multiply.
$= 5$ Simplify.

b. $f(-2)$

$f(-2) = 3(-2) - 4$ Replace x with -2.
$= -6 - 4$ Multiply.
$= -10$ Simplify.

Exercises

If $f(x) = 2x - 4$ and $g(x) = x^2 - 4x$, find each value.

1. $f(4)$
2. $g(2)$
3. $f(-5)$

4. $g(-3)$
5. $f(0)$
6. $g(0)$

7. $f(3) - 1$
8. $f\left(\dfrac{1}{4}\right)$
9. $g\left(\dfrac{1}{4}\right)$

10. $f(a^2)$
11. $f(k + 1)$
12. $g(2c)$

13. $f(3x)$
14. $f(2) + 3$
15. $g(-4)$

NAME _____ DATE _____ PERIOD _____

3-3 Study Guide and Intervention

Linear Functions

Identify Linear Equations A **linear equation** is an equation that can be written in the form $Ax + By = C$. This is called the **standard form** of a linear equation.

Standard Form of a Linear Equation	$Ax + By = C$, where $A \geq 0$, A and B are not both zero, and A, B, and C are integers whose GCF is 1.

Example 1 Determine whether $y = 6 - 3x$ is a linear equation. If so, write the equation in standard form.

First rewrite the equation so both variables are on the same side of the equation.

$y = 6 - 3x$ Original equation.
$y + 3x = 6 - 3x + 3x$ Add 3x to each side.
$3x + y = 6$ Simplify.

The equation is now in standard form, with $A = 3$, $B = 1$ and $C = 6$. This is a linear equation.

Example 2 Determine whether $3xy + y = 4 + 2x$ is a linear equation. If so, write the equation in standard form.

Since the term $3xy$ has two variables, the equation cannot be written in the form $Ax + By = C$. Therefore, this is not a linear equation.

Exercises

Determine whether each equation is a linear equation. If so, write the equation in standard form.

1. $2x = 4y$

2. $6 + y = 8$

3. $4x - 2y = -1$

4. $3xy + 8 = 4y$

5. $3x - 4 = 12$

6. $y = x^2 + 7$

7. $y - 4x = 9$

8. $x + 8 = 0$

9. $-2x + 3 = 4y$

10. $2 + \frac{1}{2}x = y$

11. $\frac{1}{4}y = 12 - 4x$

12. $3xy - y = 8$

13. $6x + 4y - 3 = 0$

14. $yx - 2 = 8$

15. $6a - 2b = 8 + b$

16. $\frac{1}{4}x - 12y = 1$

17. $3 + x + x^2 = 0$

18. $x^2 = 2xy$

Study Guide and Intervention   Glencoe Algebra 1

3-3 Study Guide and Intervention (continued)

Linear Functions

Graph Linear Equations The graph of a linear equations represents all the solutions of the equation. An x-coordinate of the point at which a graph of an equation crosses the x-axis in an **x-intercept**. A y-coordinate of the point at which a graph crosses the y-axis is called a **y-intercept**.

Example 1 Graph the equation $y - 2x = 1$ by making a table.

Solve the equation for y.

$y - 2x = 1$ Original equation.
$y - 2x + 2x = 1 + 2x$ Add 2x to each side.
$y = 2x + 1$ Simplify.

Select five values for the domain and make a table. Then graph the ordered pairs and draw a line through the points.

x	2x + 1	y	(x, y)
-2	2(-2) + 1	-3	(-2, -3)
-1	2(-1) + 1	-1	(-1, -1)
0	2(0) + 1	1	(0, 1)
1	2(1) + 1	3	(1, 3)
2	2(2) + 1	5	(2, 5)

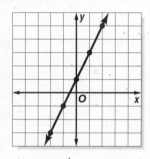

Example 2 Graph the equation $3x + 2y = 6$ by using the x-intercept and y-intercept.

To find the x-intercept, let $y = 0$ and solve for x. The x-intercept is 2. The graph intersects the x-axis at (2, 0).

To find the y-intercept, let $x = 0$ and solve for y.

The y-intercept is 3. The graph intersects the y-axis at (0, 3).

Plot the points (2, 0) and (0, 3) and draw the line through them.

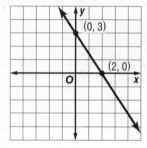

Exercises

Graph each equation by making a table.

1. $y = 2x$

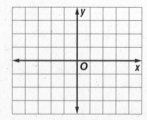

2. $x - y = -1$

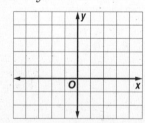

3. $x + 2y = 4$

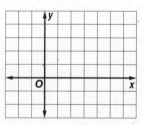

Graph each equation by using the x-intercept and y-intercept.

4. $2x + y = -2$

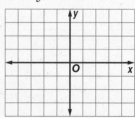

5. $3x - 6y = -3$

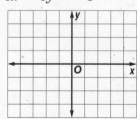

6. $-2x + y = -2$

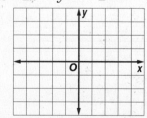

NAME _____ DATE _____ PERIOD _____

3-4 Study Guide and Intervention

Arithmetic Sequences

Recognize Arithmetic Sequences A **sequence** is a set of numbers in a specific order. If the difference between successive terms is constant, then the sequence is called an **arithmetic sequence**.

Arithmetic Sequence	a numerical pattern that increases or decreases at a constant rate or value called the **common difference**

Example 1 Determine whether the sequence 1, 3, 5, 7, 9, 11, ... is an arithmetic sequence. Justify your answer.

If possible, find the common difference between the terms. Since $3 - 1 = 2$, $5 - 3 = 2$, and so on, the common difference is 2.

Since the difference between the terms of 1, 3, 5, 7, 9, 11, ... is constant, this is an arithmetic sequence.

Example 2 Determine whether the sequence 1, 2, 4, 8, 16, 32, ... is an arithmetic sequence. Justify your answer.

If possible, find the common difference between the terms. Since $2 - 1 = 1$ and $4 - 2 = 2$, there is no common difference.

Since the difference between the terms of 1, 2, 4, 8, 16, 32, ... is not constant, this is not an arithmetic sequence.

Exercises

Determine whether each sequence is an arithmetic sequence. If it is, state the common difference.

1. 1, 5, 9, 13, 17, ...

2. 8, 4, 0, −4, −8, ...

3. 1, 3, 9, 27, 81, ...

4. 10, 15, 25, 40, 60, ...

5. −10, −5, 0, 5, 10, ...

6. 8, 6, 4, 2, 0, −2, ...

7. 4, 8, 12, 16, ...

8. 15, 12, 10, 9, ...

9. 1.1, 2.1, 3.1, 4.1, 5.1, ...

10. 8, 7, 6, 5, 4, ...

11. 0.5, 1.5, 2.5, 3.5, 4.5, ...

12. 1, 4, 16, 64, ...

13. 10, 14, 18, 22, ...

14. −3, −6, −9, −12, ...

15. 7, 0, −7, −14, ...

NAME _____ DATE _____ PERIOD _____

3-4 Study Guide and Intervention (continued)
Arithmetic Sequences

Write Arithmetic Sequences You can use the common difference of an arithmetic sequence to find the next term of the sequence. Each term after the first term is found by adding the preceding term and the common difference.

Terms of an Arithmetic Sequence	If a_1 is the first term of an arithmetic sequence with common difference d, then the sequence is a_1, $a_1 + d$, $a_1 + 2d$, $a_1 + 3d$, ….
nth Term of an Arithmetic Sequence	$a_n = a_1 + (n - 1)d$

Example 1 Find the next three terms of the arithmetic sequence 28, 32, 36, 40, ….

Find the common difference by subtracting successive terms.

28, 32, 36, 40 (+4 each)

The common difference is 4.
Add 4 to the last given term, 40, to get the next term. Continue adding 4 until the next three terms are found.

40, 44, 48, 52 (+4 each)

The next three terms are 44, 48, 52.

Example 2 Write an equation for the nth term of the sequence 12, 15, 18, 21, … .

In this sequence, a_1 is 12. Find the common difference.

12, 15, 18, 21 (+3 each)

The common difference is 3.
Use the formula for the nth term to write an equation.

$a_n = a_1 + (n - 1)d$ Formula for the nth term
$a_n = 12 + (n - 1)3$ $a_1 = 12, d = 3$
$a_n = 12 + 3n - 3$ Distributive Property
$a_n = 3n + 9$ Simplify.

The equation for the nth term is $a_n = 3n + 9$.

Exercises

Find the next three terms of each arithmetic sequence.

1. 9, 13, 17, 21, 25, …
2. 4, 0, −4, −8, −12, …
3. 29, 35, 41, 47, …

4. −10, −5, 0, 5, …
5. 2.5, 5, 7.5, 10, …
6. 3.1, 4.1, 5.1, 6.1, …

Find the nth term of each arithmetic sequence described.

7. $a_1 = 6, d = 3, n = 10$
8. $a_1 = -2, d = -3, n = 8$
9. $a_1 = 1, d = -5, n = 20$

10. $a_1 = -3, d = -2, n = 50$
11. $a_1 = -12, d = 4, n = 20$
12. $a_1 = 1, d = \frac{1}{2}, n = 11$

Write an equation for the nth term of the arithmetic sequence.

13. 1, 3, 5, 7, …
14. −1, −4, −7, −10, …
15. −4, −9, −14, −19, …

NAME _____ DATE _____ PERIOD _____

3-5 Study Guide and Intervention

Describing Number Patterns

Look for Patterns A very common problem-solving strategy is to **look for a pattern**. Arithmetic sequences follow a pattern, and other sequences can follow a pattern.

Example 1 Find the next three terms in the sequence 3, 9, 27, 81, ...

Study the pattern in the sequence.

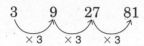

Successive terms are found by multiplying the last given term by 3.

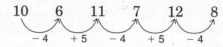

The next three terms are 243, 729, 2187.

Example 2 Find the next three terms in the sequence 10, 6, 11, 7, 12, 8, ...

Study the pattern in the sequence.

10 6 11 7 12 8
 −4 +5 −4 +5 −4

Assume that the pattern continues.

8 13 9 14
 +5 −4 +5

The next three terms are 13, 9, 14.

Exercises

1. Give the next two items for the pattern below.

Give the next three numbers in each sequence.

2. 2, 12, 72, 432, ...

3. 7, −14, 28, −56, ...

4. 0, 10, 5, 15, 10, ...

5. 0, 1, 3, 6, 10, ...

6. $x − 1, x − 2, x − 3, ...$

7. $x, \dfrac{x}{2}, \dfrac{x}{3}, \dfrac{x}{4}, ...$

NAME _____ DATE _____ PERIOD _____

3-5 Study Guide and Intervention (continued)

Describing Number Patterns

Write Equations Sometimes a pattern can lead to a general rule that can be written as an equation.

Example Suppose you purchased a number of packages of blank compact disks. If each package contains 3 compact disks, you could make a chart to show the relationship between the number of packages of compact disks and the number of disks purchased. Use x for the number of packages and y for the number of compact disks.

Make a table of ordered pairs for several points of the graph.

Number of Packages	1	2	3	4	5
Number of CDs	3	6	9	12	15

The difference in the x values is 1, and the difference in the y values is 3. This pattern shows that y is always three times x. This suggests the relation $y = 3x$. Since the relation is also a function, we can write the equation in functional notation as $f(x) = 3x$.

Exercises

1. Write an equation for the function in functional notation. Then complete the table.

x	−1	0	1	2	3	4
y	−2	2	6			

2. Write an equation for the function in functional notation. Then complete the table.

x	−2	−1	0	1	2	3
y	10	7	4			

3. Write an equation in functional notation.

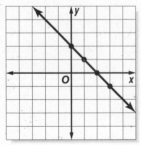

4. Write an equation in functional notation.

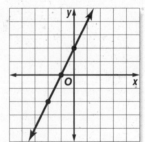

4-1 Study Guide and Intervention

Rate of Change and Slope

Rate of Change The **rate of change** tells, on average, how a quantity is changing over time. **Slope** describes a rate of change.

Example POPULATION The graph shows the population growth in China.

a. Find the rates of change for 1950–1975 and for 1975–2000.

1950–1975: $\dfrac{\text{change in population}}{\text{change in time}} = \dfrac{0.93 - 0.55}{1975 - 1950}$

$= \dfrac{0.38}{25}$ or 0.0152

1975–2000: $\dfrac{\text{change in population}}{\text{change in time}} = \dfrac{1.24 - 0.93}{2000 - 1975}$

$= \dfrac{0.31}{25}$ or 0.0124

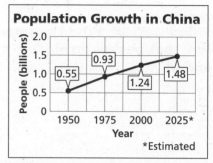

Source: United Nations Population Division

b. Explain the meaning of the slope in each case.

From 1950–1975, the growth was 0.0152 billion per year, or 15.2 million per year.
From 1975–2000, the growth was 0.0124 billion per year, or 12.4 million per year.

c. How are the different rates of change shown on the graph?

There is a greater vertical change for 1950–1975 than for 1975–2000. Therefore, the section of the graph for 1950–1975 has a steeper slope.

Exercises

LONGEVITY The graph shows the predicted life expectancy for men and women born in a given year.

1. Find the rates of change for women from 2000–2025 and 2025–2050.

2. Find the rates of change for men from 2000–2025 and 2025–2050.

3. Explain the meaning of your results in Exercises 1 and 2.

4. What pattern do you see in the increase with each 25-year period?

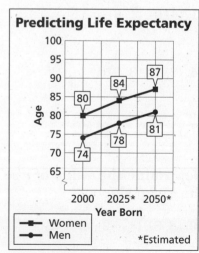

Source: USA TODAY

5. Make a prediction for the life expectancy for 2050–2075. Explain how you arrived at your prediction.

4-1 Study Guide and Intervention (continued)

Rate of Change and Slope

Find Slope

Slope of a Line	$m = \dfrac{\text{rise}}{\text{run}}$ or $m = \dfrac{y_2 - y_1}{x_2 - x_1}$, where (x_1, y_1) and (x_2, y_2) are the coordinates of any two points on a nonvertical line

Example 1 Find the slope of the line that passes through $(-3, 5)$ and $(4, -2)$.

Let $(-3, 5) = (x_1, y_1)$ and $(4, -2) = (x_2, y_2)$.

$m = \dfrac{y_2 - y_1}{x_2 - x_1}$ Slope formula

$= \dfrac{-2 - 5}{4 - (-3)}$ $y_2 = -2, y_1 = 5, x_2 = 4, x_1 = -3$

$= \dfrac{-7}{7}$ Simplify.

$= -1$

Example 2 Find the value of r so that the line through $(10, r)$ and $(3, 4)$ has a slope of $-\dfrac{2}{7}$.

$m = \dfrac{y_2 - y_1}{x_2 - x_1}$ Slope formula

$-\dfrac{2}{7} = \dfrac{4 - r}{3 - 10}$ $m = -\dfrac{2}{7}, y_2 = 4, y_1 = r, x_2 = 3, x_1 = 10$

$-\dfrac{2}{7} = \dfrac{4 - r}{-7}$ Simplify.

$-2(-7) = 7(4 - r)$ Cross multiply.

$14 = 28 - 7r$ Distributive Property

$-14 = -7r$ Subtract 28 from each side.

$2 = r$ Divide each side by -7.

Exercises

Find the slope of the line that passes through each pair of points.

1. $(4, 9), (1, 6)$
2. $(-4, -1), (-2, -5)$
3. $(-4, -1), (-4, -5)$

4. $(2, 1), (8, 9)$
5. $(14, -8), (7, -6)$
6. $(4, -3), (8, -3)$

7. $(1, -2), (6, 2)$
8. $(2, 5), (6, 2)$
9. $(4, 3.5), (-4, 3.5)$

Determine the value of r so the line that passes through each pair of points has the given slope.

10. $(6, 8), (r, -2), m = 1$
11. $(-1, -3), (7, r), m = \dfrac{3}{4}$
12. $(2, 8), (r, -4)\ m = -3$

13. $(7, -5), (6, r), m = 0$
14. $(r, 4), (7, 1), m = \dfrac{3}{4}$
15. $(7, 5), (r, 9), m = 6$

16. $(10, r), (3, 4), m = -\dfrac{2}{7}$
17. $(10, 4), (-2, r), m = -0.5$
18. $(r, 3), (7, r), m = -\dfrac{1}{5}$

NAME _____ DATE _____ PERIOD _____

4-2 Study Guide and Intervention

Slope and Direct Variation

Direct Variation A **direct variation** is described by an equation of the form $y = kx$, where $k \neq 0$. We say that y *varies directly as* x. In the equation $y = kx$, k is the **constant of variation**.

Example 1 Name the constant of variation for the equation. Then find the slope of the line that passes through the pair of points.

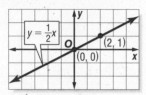

For $y = \frac{1}{2}x$, the constant of variation is $\frac{1}{2}$.

$m = \dfrac{y_2 - y_1}{x_2 - x_1}$ Slope formula

$= \dfrac{1 - 0}{2 - 0}$ $(x_1, y_1) = (0, 0), (x_2, y_2) = (2, 1)$

$= \dfrac{1}{2}$ Simplify.

The slope is $\frac{1}{2}$.

Example 2 Suppose y varies directly as x, and $y = 30$ when $x = 5$.

a. Write a direct variation equation that relates x and y.

Find the value of k.
$y = kx$ Direct variation equation
$30 = k(5)$ Replace y with 30 and x with 5.
$6 = k$ Divide each side by 5.
Therefore, the equation is $y = 6x$.

b. Use the direct variation equation to find x when $y = 18$.

$y = 6x$ Direct variation equation
$18 = 6x$ Replace y with 18.
$3 = x$ Divide each side by 6.
Therefore, $x = 3$ when $y = 18$.

Exercises

Name the constant of variation for each equation. Then determine the slope of the line that passes through each pair of points.

1.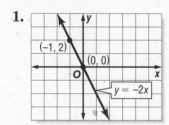

2.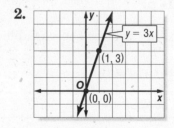

3.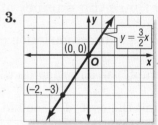

Write a direct variation equation that relates x to y. Assume that y varies directly as x. Then solve.

4. If $y = 4$ when $x = 2$, find y when $x = 16$.

5. If $y = 9$ when $x = -3$, find x when $y = 6$.

6. If $y = -4.8$ when $x = -1.6$, find x when $y = -24$.

7. If $y = \frac{1}{4}$ when $x = \frac{1}{8}$, find x when $y = \frac{3}{16}$.

NAME _____ DATE _____ PERIOD _____

4-2 Study Guide and Intervention (continued)

Slope and Direct Variation

Solve Problems The **distance formula** $d = rt$ is a direct variation equation. In the formula, distance d varies directly as time t, and the rate r is the constant of variation.

Example TRAVEL A family drove their car 225 miles in 5 hours.

a. Write a direct variation equation to find the distance traveled for any number of hours.

Use given values for d and t to find r.

$d = rt$ Original equation
$225 = r(5)$ $d = 225$ and $t = 5$
$45 = r$ Divide each side by 5.

Therefore, the direct variation equation is $d = 45t$.

b. Graph the equation.

The graph of $d = 45t$ passes through the origin with slope 45.

$m = \dfrac{45}{1}$ $\dfrac{\text{rise}}{\text{run}}$

✓ CHECK (5, 225) lies on the graph.

c. Estimate how many hours it would take the family to drive 360 miles.

$d = 45t$ Original equation
$360 = 45t$ Replace d with 360.
$t = 8$ Divide each side by 45.

Therefore, it will take 8 hours to drive 360 miles.

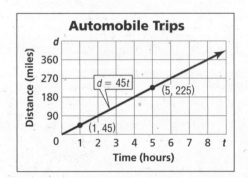

Exercises

RETAIL The total cost C of bulk jelly beans is $4.49 times the number of pounds p.

1. Write a direct variation equation that relates the variables.

2. Graph the equation on the grid at the right.

3. Find the cost of $\dfrac{3}{4}$ pound of jelly beans.

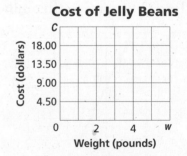

CHEMISTRY Charles's Law states that, at a constant pressure, volume of a gas V varies directly as its temperature T. A volume of 4 cubic feet of a certain gas has a temperature of 200° (absolute temperature).

4. Write a direct variation equation that relates the variables.

5. Graph the equation on the grid at the right.

6. Find the volume of the same gas at 250° (absolute temperature).

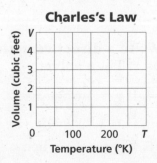

Study Guide and Intervention   50   Glencoe Algebra 1

NAME _____ DATE _____ PERIOD _____

4-3 Study Guide and Intervention

Graphing Equations in Slope-Intercept Form

Slope-Intercept Form

Slope-Intercept Form	$y = mx + b$, where m is the given slope and b is the y-intercept

Example 1 Write an equation of the line whose slope is −4 and whose y-intercept is 3.

$y = mx + b$ Slope-intercept form
$y = -4x + 3$ Replace m with −4 and b with 3.

Example 2 Graph $3x - 4y = 8$.

$3x - 4y = 8$ Original equation
$-4y = -3x + 8$ Subtract $3x$ from each side.
$\dfrac{-4y}{-4} = \dfrac{-3x + 8}{-4}$ Divide each side by −4.
$y = \dfrac{3}{4}x - 2$ Simplify.

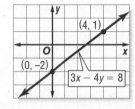

The y-intercept of $y = \dfrac{3}{4}x - 2$ is −2 and the slope is $\dfrac{3}{4}$. So graph the point $(0, -2)$. From this point, move up 3 units and right 4 units. Draw a line passing through both points.

Exercises

Write an equation of the line with the given slope and y-intercept.

1. slope: 8, y-intercept −3
2. slope: −2, y-intercept −1
3. slope: −1, y-intercept −7

Write an equation of the line shown in each graph.

4.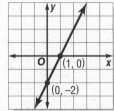
5.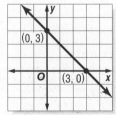
6.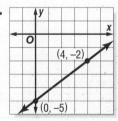

Graph each equation.

7. $y = 2x + 1$
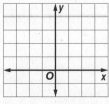

8. $y = -3x + 2$
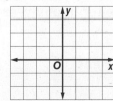

9. $y = -x - 1$
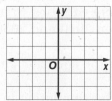

NAME _____ DATE _____ PERIOD _____

4-3 Study Guide and Intervention (continued)

Graphing Equations in Slope-Intercept Form

Model Real-World Data

Example MEDIA Since 1999, the number of music cassettes sold has decreased by an average rate of 27 million per year. There were 124 million music cassettes sold in 1999.

a. **Write a linear equation to find the average number of music cassettes sold in any year after 1999.**

The rate of change is −27 million per year. In the first year, the number of music cassettes sold was 124 million. Let N = the number of millions of music cassettes sold. Let x = the number of years after 1999. An equation is $N = -27x + 124$.

b. **Graph the equation.**

The graph of $N = -27x + 124$ is a line that passes through the point at (0, 124) and has a slope of −27.

c. **Find the approximate number of music cassettes sold in 2003.**

$N = -27x + 124$ Original equation
$N = -27(4) + 124$ Replace x with 3.
$N = 16$ Simplify.

There were about 16 million music cassettes sold in 2003.

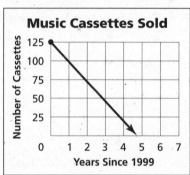
Music Cassettes Sold
Source: The World Almanac

Exercises

MUSIC In 1999, Rap and Hip-Hop music represented 9.7% of total music sales. Beween 1999 and 2003, the percent increased by about 0.7% per year.

1. Write an equation to find the percent P of recorded music and music videos sold that was Rap and Hip-Hop music for any year x between 1999 and 2003.

2. Graph the equation on the grid at the right.

3. Find the percent of recorded music and music videos sold that was Rap and Hip-Hop music in 2001.

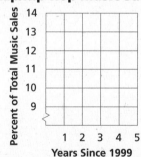
Rap/Hip-Hop Music Sales
Source: The World Almanac

POPULATION The population of the United States is projected to be 300 million by the year 2010. Between 2010 and 2050, the population is expected to increase by about 2.5 million per year.

4. Write an equation to find the population P in any year x between 2010 and 2050.

5. Graph the equation on the grid at the right.

6. Find the population in 2050.

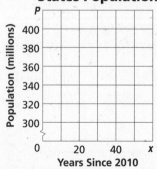
Projected United States Population
Source: The World Almanac

Study Guide and Intervention   52   Glencoe Algebra 1

4-4 Study Guide and Intervention

Writing Equations in Slope-Intercept Form

Write an Equation Given the Slope and One Point

Example 1 Write an equation of a line that passes through $(-4, 2)$ with slope 3.

The line has slope 3. To find the y-intercept, replace m with 3 and (x, y) with $(-4, 2)$ in the slope-intercept form. Then solve for b.

$y = mx + b$ Slope-intercept form
$2 = 3(-4) + b$ $m = 3, y = 2,$ and $x = -4$
$2 = -12 + b$ Multiply.
$14 = b$ Add 12 to each side.

Therefore, the equation is $y = 3x + 14$.

Example 2 Write an equation of the line that passes through $(-2, -1)$ with slope $\frac{1}{4}$.

The line has slope $\frac{1}{4}$. Replace m with $\frac{1}{4}$ and (x, y) with $(-2, -1)$ in the slope-intercept form.

$y = mx + b$ Slope-intercept form
$-1 = \frac{1}{4}(-2) + b$ $m = \frac{1}{4}, y = -1,$ and $x = -2$
$-1 = -\frac{1}{2} + b$ Multiply.
$-\frac{1}{2} = b$ Add $\frac{1}{2}$ to each side.

Therefore, the equation is $y = \frac{1}{4}x - \frac{1}{2}$.

Exercises

Write an equation of the line that passes through each point with the given slope.

1.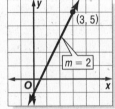
2.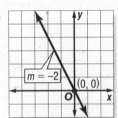
3.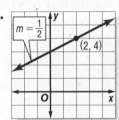

4. $(8, 2), m = -\frac{3}{4}$

5. $(-1, -3), m = 5$

6. $(4, -5), m = -\frac{1}{2}$

7. $(-5, 4), m = 0$

8. $(2, 2), m = \frac{1}{2}$

9. $(1, -4), m = -6$

10. Write an equation of a line that passes through the y-intercept -3 with slope 2.

11. Write an equation of a line that passes through the x-intercept 4 with slope -3.

12. Write an equation of a line that passes through the point $(0, 350)$ with slope $\frac{1}{5}$.

NAME _____ DATE _____ PERIOD _____

4-4 Study Guide and Intervention (continued)

Writing Equations in Slope-Intercept Form

Write an Equation Given Two Points

Example Write an equation of the line that passes through $(1, 2)$ and $(3, -2)$.

Find the slope m. To find the y-intercept, replace m with its computed value and (x, y) with $(1, 2)$ in the slope-intercept form. Then solve for b.

$m = \dfrac{y_2 - y_1}{x_2 - x_1}$ Slope formula

$m = \dfrac{-2 - 2}{3 - 1}$ $y_2 = -2, y_1 = 2, x_2 = 3, x_1 = 1$

$m = -2$ Simplify.

$y = mx + b$ Slope-intercept form

$2 = -2(1) + b$ Replace m with -2, y with 2, and x with 1.

$2 = -2 + b$ Multiply.

$4 = b$ Add 2 to each side.

Therefore, the equation is $y = -2x + 4$.

Exercises

Write an equation of the line that passes through each pair of points.

1.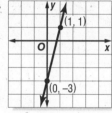
2.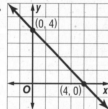
3.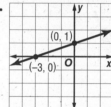

4. $(-1, 6), (7, -10)$ 5. $(0, 2), (1, 7)$ 6. $(6, -25), (-1, 3)$

7. $(-2, -1), (2, 11)$ 8. $(10, -1), (4, 2)$ 9. $(-14, -2), (7, 7)$

10. Write an equation of a line that passes through the x-intercept 4 and y-intercept -2.

11. Write an equation of a line that passes through the x-intercept -3 and y-intercept 5.

12. Write an equation of a line that passes through $(0, 16)$ and $(-10, 0)$.

NAME _____ DATE _____ PERIOD _____

4-5 Study Guide and Intervention
Writing Equations in Point-Slope Form

Point-Slope Form

| Point-Slope Form | $y - y_1 = m(x - x_1)$, where (x_1, y_1) is a given point on a nonvertical line and m is the slope of the line |

Example 1 Write the point-slope form of an equation for a line that passes through (6, 1) and has a slope of $-\frac{5}{2}$.

$y - y_1 = m(x - x_1)$ Point-slope form
$y - 1 = -\frac{5}{2}(x - 6)$ $m = -\frac{5}{2}$; $(x_1, y_1) = (6, 1)$

Therefore, the equation is $y - 1 = -\frac{5}{2}(x - 6)$.

Example 2 Write the point-slope form of an equation for a horizontal line that passes through (4, −1).

$y - y_1 = m(x - x_1)$ Point-slope form
$y - (-1) = 0(x - 4)$ $m = 0$; $(x_1, y_1) = (4, -1)$
$y + 1 = 0$ Simplify.

Therefore, the equation is $y + 1 = 0$.

Exercises

Write the point-slope form of an equation for a line that passes through each point with the given slope.

1.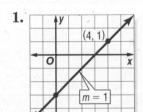

2.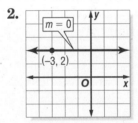

3.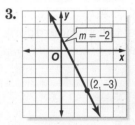

4. (2, 1), $m = 4$

5. (−7, 2), $m = 6$

6. (8, 3), $m = 1$

7. (−6, 7), $m = 0$

8. (4, 9), $m = \frac{3}{4}$

9. (−4, −5), $m = -\frac{1}{2}$

10. Write the point-slope form of an equation for the horizontal line that passes through (4, −2).

11. Write the point-slope form of an equation for the horizontal line that passes through (−5, 6).

12. Write the point-slope form of an equation for the horizontal line that passes through (5, 0).

Study Guide and Intervention   Glencoe Algebra 1

NAME _____ DATE _____ PERIOD _____

4-5 Study Guide and Intervention (continued)
Writing Equations in Point-Slope Form

Forms of Linear Equations

Slope-Intercept Form	$y = mx + b$	m = slope; b = y-intercept
Point-Slope Form	$y - y_1 = m(x - x_1)$	m = slope; (x_1, y_1) is a given point.
Standard Form	$Ax + By = C$	A and B are not both zero. Usually A is nonnegative and A, B, and C are integers whose greatest common factor is 1.

Example 1 Write $y + 5 = \frac{2}{3}(x - 6)$ in standard form.

$y + 5 = \frac{2}{3}(x - 6)$ Original equation

$3(y + 5) = 3\left(\frac{2}{3}\right)(x - 6)$ Multiply each side by 3.

$3y + 15 = 2(x - 6)$ Distributive Property
$3y + 15 = 2x - 12$ Distributive Property
$3y = 2x - 27$ Subtract 15 from each side.
$-2x + 3y = -27$ Add $-2x$ to each side.
$2x - 3y = 27$ Multiply each side by -1.

Therefore, the standard form of the equation is $2x - 3y = 27$.

Example 2 Write $y - 2 = -\frac{1}{4}(x - 8)$ in slope-intercept form.

$y - 2 = -\frac{1}{4}(x - 8)$ Original equation

$y - 2 = -\frac{1}{4}x + 2$ Distributive Property

$y = -\frac{1}{4}x + 4$ Add 2 to each side.

Therefore, the slope-intercept form of the equation is $y = -\frac{1}{4}x + 4$.

Exercises

Write each equation in standard form.

1. $y + 2 = -3(x - 1)$

2. $y - 1 = -\frac{1}{3}(x - 6)$

3. $y + 2 = \frac{2}{3}(x - 9)$

4. $y + 3 = -(x - 5)$

5. $y - 4 = \frac{5}{3}(x + 3)$

6. $y + 4 = -\frac{2}{5}(x - 1)$

Write each equation in slope-intercept form.

7. $y + 4 = 4(x - 2)$

8. $y - 5 = \frac{1}{3}(x - 6)$

9. $y - 8 = -\frac{1}{4}(x + 8)$

10. $y - 6 = 3\left(x - \frac{1}{3}\right)$

11. $y + 4 = -2(x + 5)$

12. $y + \frac{5}{3} = \frac{1}{2}(x - 2)$

NAME _____ DATE _____ PERIOD _____

4-6 Study Guide and Intervention

Statistics: Scatter Plots and Lines of Fit

Interpret Points on a Scatter Plot A **scatter plot** is a graph in which two sets of data are plotted as ordered pairs in a coordinate plane. If y increases as x increases, there is a **positive correlation** between x and y. If y decreases as x increases, there is a **negative correlation** between x and y. If x and y are not related, there is **no correlation**.

Example EARNINGS The graph at the right shows the amount of money Carmen earned each week and the amount she deposited in her savings account that same week. Determine whether the graph shows a positive correlation, a negative correlation, or no correlation. If there is a positive or negative correlation, describe its meaning in the situation.

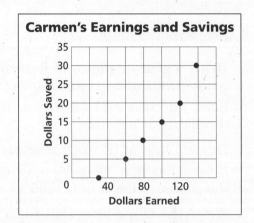

The graph shows a positive correlation. The more Carmen earns, the more she saves.

Exercises

Determine whether each graph shows a positive correlation, a negative correlation, or no correlation. If there is a positive correlation, describe it.

1. Average Weekly Work Hours in U.S.

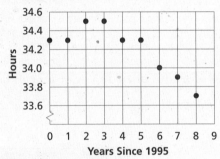

Source: *The World Almanac*

2. Average Jogging Speed

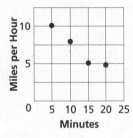

3. Average Hourly Earnings in U.S.

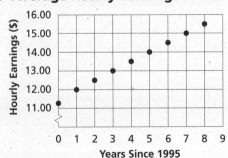

Source: *The World Almanac*

4. U.S. Imports from Mexico

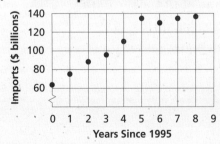

Source: *The World Almanac*

NAME _____ DATE _____ PERIOD _____

4-6 Study Guide and Intervention (continued)

Statistics: Scatter Plots and Lines of Fit

Lines of Fit

Example The table below shows the number of students per computer in United States public schools for certain school years from 1990 to 2000.

Year	1990	1992	1994	1996	1998	2000	2002
Students per Computer	22	18	14	10	6.1	5.4	4.9

a. **Draw a scatter plot and determine what relationship exists, if any.**

Since y decreases as x increases, the correlation is negative.

b. **Draw a line of fit for the scatter plot.**

Draw a line that passes close to most of the points. A line of fit is shown.

c. **Write the slope-intercept form of an equation for the line of fit.**

The line of fit shown passes through (1993, 16) and (1999, 5.7). Find the slope.

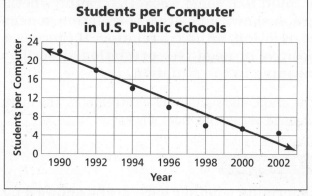

Source: The World Almanac

$$m = \frac{5.7 - 16}{1999 - 1993}$$

$m = -1.7$

Find b in $y = -1.7x + b$.

$16 = -1.7 \cdot 1993 + b$

$3404 = b$

Therefore, an equation of a line of fit is $y = -1.7x + 3404$.

Exercises

Refer to the table for Exercises 1–3.

1. Draw a scatter plot.
2. Draw a line of fit for the data.
3. Write the slope-intercept form of an equation for the line of fit.

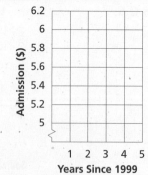

Source: U.S. Census Bureau

Years Since 1999	Admission (dollars)
0	$5.08
1	$5.39
2	$5.66
3	$5.81
4	$6.03

NAME _____ DATE _____ PERIOD _____

4-7 Study Guide and Intervention

Geometry: Parallel and Perpendicular Lines

Parallel Lines Two nonvertical lines are **parallel** if they have the same slope. All vertical lines are parallel.

Example Write the slope-intercept form for an equation of the line that passes through $(-1, 6)$ and is parallel to the graph of $y = 2x + 12$.

A line parallel to $y = 2x + 12$ has the same slope, 2. Replace m with 2 and (x_1, y_1) with $(-1, 6)$ in the point-slope form.

$y - y_1 = m(x - x_1)$ Point-slope form
$y - 6 = 2(x - (-1))$ $m = 2; (x_1, y_1) = (-1, 6)$
$y - 6 = 2(x + 1)$ Simplify.
$y - 6 = 2x + 2$ Distributive Property
$y = 2x + 8$ Slope-intercept form

Therefore, the equation is $y = 2x + 8$.

Exercises

Write the slope-intercept form for an equation of the line that passes through the given point and is parallel to the graph of each equation.

1.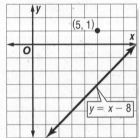
2.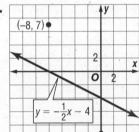
3.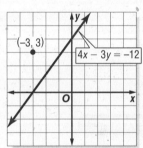

4. $(-2, 2)$, $y = 4x - 2$

5. $(6, 4)$, $y = \frac{1}{3}x + 1$

6. $(4, -2)$, $y = -2x + 3$

7. $(-2, 4)$, $y = -3x + 10$

8. $(-1, 6)$, $3x + y = 12$

9. $(4, -6)$, $x + 2y = 5$

10. Find an equation of the line that has a y-intercept of 2 that is parallel to the graph of the line $4x + 2y = 8$.

11. Find an equation of the line that has a y-intercept of -1 that is parallel to the graph of the line $x - 3y = 6$.

12. Find an equation of the line that has a y-intercept of -4 that is parallel to the graph of the line $y = 6$.

Study Guide and Intervention   Glencoe Algebra 1

NAME _____ DATE _____ PERIOD _____

4-7 Study Guide and Intervention (continued)
Geometry: Parallel and Perpendicular Lines

Perpendicular Lines Two non-vertical lines are **perpendicular** if their slopes are negative reciprocals of each other. Vertical and horizontal lines are perpendicular.

Example Write the slope-intercept form for an equation that passes through $(-4, 2)$ and is perpendicular to the graph of $2x - 3y = 9$.

Find the slope of $2x - 3y = 9$.

$2x - 3y = 9$ Original equation
$-3y = -2x + 9$ Subtract 2x from each side.
$y = \frac{2}{3}x - 3$ Divide each side by -3.

The slope of $y = \frac{2}{3}x - 3$ is $\frac{2}{3}$. So, the slope of the line passing through $(-4, 2)$ that is perpendicular to this line is the negative reciprocal of $\frac{2}{3}$, or $-\frac{3}{2}$.

Use the point-slope form to find the equation.

$y - y_1 = m(x - x_1)$ Point-slope form
$y - 2 = -\frac{3}{2}(x - (-4))$ $m = -\frac{3}{2}$; $(x_1, y_1) = (-4, 2)$
$y - 2 = -\frac{3}{2}(x + 4)$ Simplify.
$y - 2 = -\frac{3}{2}x - 6$ Distributive Property
$y = -\frac{3}{2}x - 4$ Slope-intercept form

Exercises

Write the slope-intercept form for an equation of the line that passes through the given point and is perpendicular to the graph of each equation.

1. $(4, 2), y = \frac{1}{2}x + 1$ 2. $(2, -3), y = -\frac{2}{3}x + 4$ 3. $(6, 4), y = 7x + 1$

4. $(-8, -7), y = -x - 8$ 5. $(6, -2), y = -3x - 6$ 6. $(-5, -1), y = \frac{5}{2}x - 3$

7. $(-9, -5), y = -3x - 1$ 8. $(-1, 3), 2x + 4y = 12$ 9. $(6, -6), 3x - y = 6$

10. Find an equation of the line that has a y-intercept of -2 and is perpendicular to the graph of the line $x - 2y = 5$.

11. Find an equation of the line that has a y-intercept of 5 and is perpendicular to the graph of the line $4x + 3y = 8$.

NAME _____ DATE _____ PERIOD _____

5-1 Study Guide and Intervention

Graphing Systems of Equations

Number of Solutions Two or more linear equations involving the same variables form a **system of equations**. A solution of the system of equations is an ordered pair of numbers that satisfies both equations. The table below summarizes information about systems of linear equations.

Graph of a System	intersecting lines	same line	parallel lines
	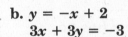		
Number of Solutions	exactly one solution	infinitely many solutions	no solution
Terminology	consistent and independent	consistent and dependent	inconsistent

Example Use the graph at the right to determine whether the system has *no* solution, *one* solution, or *infinitely many* solutions.

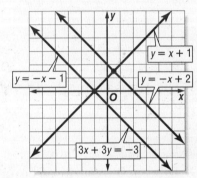

a. $y = -x + 2$
   $y = x + 1$

   Since the graphs of $y = -x + 2$ and $y = x + 1$ intersect, there is one solution.

b. $y = -x + 2$
   $3x + 3y = -3$

   Since the graphs of $y = -x + 2$ and $3x + 3y = -3$ are parallel, there are no solutions.

c. $3x + 3y = -3$
   $y = -x - 1$

   Since the graphs of $3x + 3y = -3$ and $y = -x - 1$ coincide, there are infinitely many solutions.

Exercises

Use the graph at the right to determine whether each system has *no* solution, *one* solution, or *infinitely many* solutions.

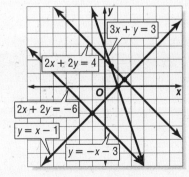

1. $y = -x - 3$
   $y = x - 1$

2. $2x + 2y = -6$
   $y = -x - 3$

3. $y = -x - 3$
   $2x + 2y = 4$

4. $2x + 2y = -6$
   $3x + y = 3$

Study Guide and Intervention   Glencoe Algebra 1

NAME _____ DATE _____ PERIOD _____

5-1 Study Guide and Intervention *(continued)*

Graphing Systems of Equations

Solve by Graphing One method of solving a system of equations is to graph the equations on the same coordinate plane.

Example Graph each system of equations. Then determine whether the system has *no solution*, *one solution*, or *infinitely many* solutions. If the system has one solution, name it.

a. $x + y = 2$
   $x - y = 4$

The graphs intersect. Therefore, there is one solution. The point $(3, -1)$ seems to lie on both lines. Check this estimate by replacing x with 3 and y with -1 in each equation.

$x + y = 2$
$3 + (-1) = 2$ ✓
$x - y = 4$
$3 - (-1) = 3 + 1$ or 4 ✓

The solution is $(3, -1)$.

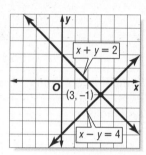

b. $y = 2x + 1$
   $2y = 4x + 2$

The graphs coincide. Therefore there are infinitely many solutions.

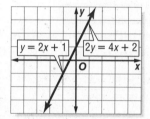

Exercises

Graph each system of equations. Then determine whether the system has *no* solution, *one* solution, or *infinitely many* solutions. If the system has one solution, name it.

1. $y = -2$
   $3x - y = -1$

2. $x = 2$
   $2x + y = 1$

3. $y = \frac{1}{2}x$
   $x + y = 3$

4. $2x + y = 6$
   $2x - y = -2$

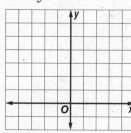

5. $3x + 2y = 6$
   $3x + 2y = -4$

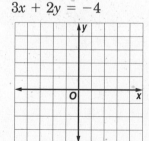

6. $2y = -4x + 4$
   $y = -2x + 2$

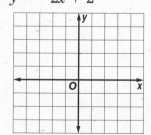

Study Guide and Intervention

NAME _____ DATE _____ PERIOD _____

5-2 Study Guide and Intervention
Substitution

Substitution One method of solving systems of equations is **substitution**.

Example 1 Use substitution to solve the system of equations.
$y = 2x$
$4x - y = -4$

Substitute $2x$ for y in the second equation.

$4x - y = -4$	Second equation
$4x - 2x = -4$	$y = 2x$
$2x = -4$	Combine like terms.
$x = -2$	Divide each side by 2 and simplify.

Use $y = 2x$ to find the value of y.

$y = 2x$	First equation
$y = 2(-2)$	$x = -2$
$y = -4$	Simplify.

The solution is $(-2, -4)$.

Example 2 Solve for one variable, then substitute.
$x + 3y = 7$
$2x - 4y = -6$

Solve the first equation for x since the coefficient of x is 1.

$x + 3y = 7$	First equation
$x + 3y - 3y = 7 - 3y$	Subtract $3y$ from each side.
$x = 7 - 3y$	Simplify.

Find the value of y by substituting $7 - 3y$ for x in the second equation.

$2x - 4y = -6$	Second equation
$2(7 - 3y) - 4y = -6$	$x = 7 - 3y$
$14 - 6y - 4y = -6$	Distributive Property
$14 - 10y = -6$	Combine like terms.
$14 - 10y - 14 = -6 - 14$	Subtract 14 from each side.
$-10y = -20$	Simplify.
$y = 2$	Divide each side by -10 and simplify.

Use $y = 2$ to find the value of x.
$x = 7 - 3y$
$x = 7 - 3(2)$
$x = 1$
The solution is $(1, 2)$.

Exercises

Use substitution to solve each system of equations. If the system does *not* have exactly one solution, state whether it has *no* solution or *infinitely many* solutions.

1. $y = 4x$
   $3x - y = 1$

2. $x = 2y$
   $y = x - 2$

3. $x = 2y - 3$
   $x = 2y + 4$

4. $x - 2y = -1$
   $3y = x + 4$

5. $c - 4d = 1$
   $2c - 8d = 2$

6. $x + 2y = 0$
   $3x + 4y = 4$

7. $2b = 6a - 14$
   $3a - b = 7$

8. $x + y = 16$
   $2y = -2x + 2$

9. $y = -x + 3$
   $2y + 2x = 4$

10. $x = 2y$
    $0.25x + 0.5y = 10$

11. $x - 2y = -5$
    $x + 2y = -1$

12. $-0.2x + y = 0.5$
    $0.4x + y = 1.1$

5-2 Study Guide and Intervention (continued)

Substitution

Real-World Problems Substitution can also be used to solve real-world problems involving systems of equations. It may be helpful to use tables, charts, diagrams, or graphs to help you organize data.

Example CHEMISTRY How much of a 10% saline solution should be mixed with a 20% saline solution to obtain 1000 milliliters of a 12% saline solution?

Let s = the number of milliliters of 10% saline solution.
Let t = the number of milliliters of 20% saline solution.
Use a table to organize the information.

	10% saline	20% saline	12% saline
Total milliliters	s	t	1000
Milliliters of saline	$0.10s$	$0.20t$	$0.12(1000)$

Write a system of equations.
$s + t = 1000$
$0.10s + 0.20t = 0.12(1000)$
Use substitution to solve this system.

$s + t = 1000$	First equation
$s = 1000 - t$	Solve for s.
$0.10s + 0.20t = 0.12(1000)$	Second equation
$0.10(1000 - t) + 0.20t = 0.12(1000)$	$s = 1000 - t$
$100 - 0.10t + 0.20t = 0.12(1000)$	Distributive Property
$100 + 0.10t = 0.12(1000)$	Combine like terms.
$0.10t = 20$	Simplify.
$\dfrac{0.10t}{0.10} = \dfrac{20}{0.10}$	Divide each side by 0.10.
$t = 200$	Simplify.
$s + t = 1000$	First equation
$s + 200 = 1000$	$t = 200$
$s = 800$	Solve for s.

800 milliliters of 10% solution and 200 milliliters of 20% solution should be used.

Exercises

1. **SPORTS** At the end of the 2000-2001 football season, 31 Super Bowl games had been played with the current two football leagues, the American Football Conference (AFC) and the National Football Conference (NFC). The NFC won five more games than the AFC. How many games did each conference win? *Source: New York Times Almanac*

2. **CHEMISTRY** A lab needs to make 100 gallons of an 18% acid solution by mixing a 12% acid solution with a 20% solution. How many gallons of each solution are needed?

3. **GEOMETRY** The perimeter of a triangle is 24 inches. The longest side is 4 inches longer than the shortest side, and the shortest side is three-fourths the length of the middle side. Find the length of each side of the triangle.

NAME _____ DATE _____ PERIOD _____

5-3 Study Guide and Intervention

Elimination Using Addition and Subtraction

Elimination Using Addition In systems of equations in which the coefficients of the x or y terms are additive inverses, solve the system by adding the equations. Because one of the variables is eliminated, this method is called **elimination**.

Example 1 Use addition to solve the system of equations.
$x - 3y = 7$
$3x + 3y = 9$

Write the equations in column form and add to eliminate y.
$x - 3y = 7$
$(+)\,3x + 3y = 9$
$4x = 16$

Solve for x.
$\dfrac{4x}{4} = \dfrac{16}{4}$
$x = 4$

Substitute 4 for x in either equation and solve for y.
$4 - 3y = 7$
$4 - 3y - 4 = 7 - 4$
$-3y = 3$
$\dfrac{-3y}{-3} = \dfrac{3}{-3}$
$y = -1$

The solution is $(4, -1)$.

Example 2 The sum of two numbers is 70 and their difference is 24. Find the numbers.

Let x represent one number and y represent the other number.
$x + y = 70$
$(+)\,x - y = 24$
$2x = 94$
$\dfrac{2x}{2} = \dfrac{94}{2}$
$x = 47$

Substitute 47 for x in either equation.
$47 + y = 70$
$47 + y - 47 = 70 - 47$
$y = 23$

The numbers are 47 and 23.

Exercises

Use elimination to solve each system of equations.

1. $x + y = -4$
   $x - y = 2$

2. $2m - 3n = 14$
   $m + 3n = -11$

3. $3a - b = -9$
   $-3a - 2b = 0$

4. $-3x - 4y = -1$
   $3x - y = -4$

5. $3c + d = 4$
   $2c - d = 6$

6. $-2x + 2y = 9$
   $2x - y = -6$

7. $2x + 2y = -2$
   $3x - 2y = 12$

8. $4x - 2y = -1$
   $-4x + 4y = -2$

9. $x - y = 2$
   $x + y = -3$

10. $2x - 3y = 12$
    $4x + 3y = 24$

11. $-0.2x + y = 0.5$
    $0.2x + 2y = 1.6$

12. $0.1x + 0.3y = 0.9$
    $0.1x - 0.3y = 0.2$

13. Rema is older than Ken. The difference of their ages is 12 and the sum of their ages is 50. Find the age of each.

14. The sum of the digits of a two-digit number is 12. The difference of the digits is 2. Find the number if the units digit is larger than the tens digit.

5-3 Study Guide and Intervention (continued)
Elimination Using Addition and Subtraction

Elimination Using Subtraction In systems of equations where the coefficients of the x or y terms are the same, solve the system by subtracting the equations.

Example Use subtraction to solve the system of equations.
$2x - 3y = 11$
$5x - 3y = 14$

$$\begin{aligned} 2x - 3y &= 11 \\ (-)\ 5x - 3y &= 14 \\ \hline -3x &= -3 \end{aligned}$$ Write the equations in column form and subtract.

Subtract the two equations. y is eliminated.

$\dfrac{-3x}{-3} = \dfrac{-3}{-3}$ Divide each side by -3.

$x = 1$ Simplify.

$2(1) - 3y = 11$ Substitute 1 for x in either equation.
$2 - 3y = 11$ Simplify.
$2 - 3y - 2 = 11 - 2$ Subtract 2 from each side.
$-3y = 9$ Simplify.
$\dfrac{-3y}{-3} = \dfrac{9}{-3}$ Divide each side by -3.
$y = -3$ Simplify.

The solution is $(1, -3)$.

Exercises

Use elimination to solve each system of equations.

1. $6x + 5y = 4$
   $6x - 7y = -20$

2. $3m - 4n = -14$
   $3m + 2n = -2$

3. $3a + b = 1$
   $a + b = 3$

4. $-3x - 4y = -23$
   $-3x + y = 2$

5. $c - 3d = 11$
   $2c - 3d = 16$

6. $x - 2y = 6$
   $x + y = 3$

7. $2a - 3b = -13$
   $2a + 2b = 7$

8. $4x + 2y = 6$
   $4x + 4y = 10$

9. $5s - t = 6$
   $5s + 2t = 3$

10. $6x - 3y = 12$
    $4x - 3y = 24$

11. $x + 2y = 3.5$
    $x - 3y = -9$

12. $0.2x + y = 0.7$
    $0.2x + 2y = 1.2$

13. The sum of two numbers is 70. One number is ten more than twice the other number. Find the numbers.

14. **GEOMETRY** Two angles are supplementary. The measure of one angle is 10° more than three times the other. Find the measure of each angle.

5-4 Study Guide and Intervention

Elimination Using Multiplication

Elimination Using Multiplication Some systems of equations cannot be solved simply by adding or subtracting the equations. In such cases, one or both equations must first be multiplied by a number before the system can be solved by elimination.

Example 1 Use elimination to solve the system of equations.
$x + 10y = 3$
$4x + 5y = 5$

If you multiply the second equation by -2, you can eliminate the y terms.

$$\begin{array}{r} x + 10y = 3 \\ (+)\,-8x - 10y = -10 \\ \hline -7x = -7 \end{array}$$

$$\frac{-7x}{-7} = \frac{-7}{-7}$$
$$x = 1$$

Substitute 1 for x in either equation.
$1 + 10y = 3$
$1 + 10y - 1 = 3 - 1$
$10y = 2$
$\frac{10y}{10} = \frac{2}{10}$
$y = \frac{1}{5}$

The solution is $\left(1, \frac{1}{5}\right)$.

Example 2 Use elimination to solve the system of equations.
$3x - 2y = -7$
$2x - 5y = 10$

If you multiply the first equation by 2 and the second equation by -3, you can eliminate the x terms.

$$\begin{array}{r} 6x - 4y = -14 \\ (+)\,-6x + 15y = -30 \\ \hline 11y = -44 \end{array}$$

$$\frac{11y}{11} = \frac{-44}{11}$$
$$y = -4$$

Substitute -4 for y in either equation.
$3x - 2(-4) = -7$
$3x + 8 = -7$
$3x + 8 - 8 = -7 - 8$
$3x = -15$
$\frac{3x}{3} = \frac{-15}{3}$
$x = -5$

The solution is $(-5, -4)$.

Exercises

Use elimination to solve each system of equations.

1. $2x + 3y = 6$
   $x + 2y = 5$

2. $2m + 3n = 4$
   $-m + 2n = 5$

3. $3a - b = 2$
   $a + 2b = 3$

4. $4x + 5y = 6$
   $6x - 7y = -20$

5. $4c - 3d = 22$
   $2c - d = 10$

6. $3x - 4y = -4$
   $x + 3y = -10$

7. $4s - t = 9$
   $5s + 2t = 8$

8. $4a - 3b = -8$
   $2a + 2b = 3$

9. $2x + 2y = 5$
   $4x - 4y = 10$

10. $6x - 4y = -8$
    $4x + 2y = -3$

11. $4x + 2y = -5$
    $-2x - 4y = 1$

12. $2x + y = 3.5$
    $-x + 2y = 2.5$

13. **GARDENING** The length of Sally's garden is 4 meters greater than 3 times the width. The perimeter of her garden is 72 meters. What are the dimensions of Sally's garden?

14. Anita is $4\frac{1}{2}$ years older than Basilio. Three times Anita's age added to six times Basilio's age is 36. How old are Anita and Basilio?

NAME _____ DATE _____ PERIOD _____

5-4 Study Guide and Intervention (continued)
Elimination Using Multiplication

Determine the Best Method The methods to use for solving systems of linear equations are summarized in the table below.

Method	The Best Time to Use
Graphing	to estimate the solution, since graphing usually does not give an exact solution
Substitution	if one of the variables in either equation has a coefficient of 1 or −1
Elimination Using Addition	if one of the variables has opposite coefficients in the two equations
Elimination Using Subtraction	if one of the variables has the same coefficient in the two equations
Elimination Using Multiplication	if none of the coefficients are 1 or −1 and neither of the variables can be eliminated by simply adding or subtracting the equations

Example
Determine the best method to solve the system of equations. Then solve the system.

$6x + 2y = 20$
$-2x + 4y = -16$

Since the coefficients of x will be additive inverses of each other if you multiply the second equation by 3, use elimination.

$6x + 2y = 20$
$(+) -6x + 12y = -48$ Multiply the second equation by 3.
$\overline{}$
$14y = -28$ Add the two equations. x is eliminated.
$\dfrac{14y}{14} = \dfrac{-28}{14}$ Divide each side by 14.
$y = -2$ Simplify.

$6x + 2(-2) = 20$ Substitute −2 for y in either equation.
$6x - 4 = 20$ Simplify.
$6x - 4 + 4 = 20 + 4$ Add 4 to each side.
$6x = 24$ Simplify.
$\dfrac{6x}{6} = \dfrac{24}{6}$ Divide each side by 6.
$x = 4$ Simplify.

The solution is $(4, -2)$.

Exercises
Determine the best method to solve each system of equations. Then solve the system.

1. $x + 2y = 3$
   $x + y = 1$

2. $m + 6n = -8$
   $m = 2n + 8$

3. $a - b = 6$
   $a = 2b + 7$

4. $4x + y = 15$
   $-x - 3y = -12$

5. $3c - d = 14$
   $c - d = 2$

6. $x + 2y = -9$
   $y = 4x$

7. $4x = 2y - 10$
   $x + 2y = 5$

8. $x = -2y$
   $4x + 4y = -10$

9. $2s - 3t = 42$
   $3s + 2t = 24$

10. $4a - 4b = -10$
    $2a + 4b = -2$

11. $4x + 10y = -6$
    $-2x - 10y = 2$

12. $2x = y - 3$
    $-x + y = 0$

Study Guide and Intervention   Glencoe Algebra 1

NAME _____ DATE _____ PERIOD _____

5-5 Study Guide and Intervention

Applying Systems of Linear Equations

DETERMINE THE BEST METHOD You have learned five methods for solving systems of linear equations: graphing, substitution, elimination using addition, elimination using subtraction, and elimination using multiplication. For an exact solution, an algebraic method is best.

Example At a baseball game, Henry bought 3 hotdogs and a bag of chips for $14. Scott bought 2 hotdogs and a bag of chips for $10. The hotdogs and chips were all the same price, so the following system of equations can be used to represent the situation. Determine the best method to solving the system of equations. Then solve the system.

$3x + y = 14$
$2x + y = 10$

- Since neither the coefficients of x nor the coefficients of y are additive inverses, you cannot use elimination using addition.
- Since the coefficient of y in both equations is 1, you can use elimination using subtraction. You could also use the substitution method or elimination using multiplication.

The following solution uses elimination by subtraction to solve this system.

$3x +$	$y =$	14	Write the equations in column form and subtract.
$(-)\,2x + (-)\,y = (-)10$		
x	$=$	4	The variable y is eliminated.
$3(4) +$	$y =$	14	Substitute the value for x back into the first equation.
	$y =$	2	Solve for y.

This means that hot dogs cost $4 each and a bag of chips costs $2.

Exercises

Determine the best method to solve each system of equations. Then solve the system.

1. $5x + 3y = 16$
   $3x - 5y = -4$

2. $3x - 5y = 7$
   $2x + 5y = 13$

3. $y + 3x = 24$
   $5x - y = 8$

4. $-11x - 10y = 17$
   $5x - 7y = 50$

NAME _____ DATE _____ PERIOD _____

5-5 Study Guide and Intervention (continued)
Applying Systems of Linear Equations

APPLY SYSTEMS OF LINEAR EQUATIONS When applying systems of linear equations to problem situations, it is important to analyze each solution in the context of the situation.

Example BUSINESS A T-shirt printing company sells T-shirt for $15 each. The company has a fixed cost for the machine used to print the T-shirts and an additional cost per T-shirt. Use the table to estimate the number of T-shirts the company must sell in order for the income equal to expenses.

T-shirt Printing Cost	
Printing machine	$3000.00
blank T-shirt	$5.00

Explore You know the initial income and the initial expense and the rates of change of each quantity with each T-shirt sold.

Plan Write an equation to represent the income and the expenses. Then solve to find how many T-shirts need to be sold for both values to be equal.

Solve Let x = the number of T-shirts sold and let y = the total amount.

	total amount	initial amount	rate of change times number of T-shirts sold
income	$y =$	0 +	$15x$
expenses	$y =$	3000 +	$5x$

You can use substitution to solve this system.

$y = 15x$ The first equation.

$15x = 3000 + 5x$ Substitute the value for y into the second equation.

$10x = 3000$ Subtract 10x from each side and simplify.

$x = 300$ Divide each side by 10 and simplify.

This means that if 300 T-shirts are sold, the income and expenses of the T-shirt company are equal.

Check Does this solution make sense in the context of the problem? After selling 100 T-shirts, the income would be about 100 × $15 or $1500. The costs would be about $3000 + 100 × $5 or $3500.

Exercises

Refer to the example above. If the costs of the T-shirt company change to the given values and the selling price remains the same, determine the number of T-shirts the company must sell in order for income to equal expenses.

1. printing machine: $5000.00;
   T-shirt: $10.00 each

2. printing machine: $2100.00;
   T-shirt: $8.00 each

3. printing machine: $8800.00;
   T-shirt: $4.00 each

4. printing machine: $1200.00;
   T-shirt: $12.00 each

NAME _____ DATE _____ PERIOD _____

6-1 Study Guide and Intervention

Solving Inequalities by Addition and Subtraction

Solve Inequalities by Addition Addition can be used to solve inequalities. If any number is added to each side of a true inequality, the resulting inequality is also true.

Addition Property of Inequalities	For all numbers a, b, and c, if $a > b$, then $a + c > b + c$, and if $a < b$, then $a + c < b + c$.

The property is also true when $>$ and $<$ are replaced with $\geq$ and $\leq$.

Example 1 Solve $x - 8 \leq -6$. Then graph it on a number line.

$x - 8 \leq -6$ Original inequality
$x - 8 + 8 \leq -6 + 8$ Add 8 to each side.
$x \leq 2$ Simplify.

The solution in set-builder notation is $\{x \mid x \leq 2\}$.
Number line graph:

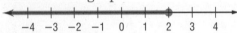

Example 2 Solve $4 - 2a > -a$. Then graph it on a number line.

$4 - 2a > -a$ Original inequality
$4 - 2a + 2a > -a + 2a$ Add 2a to each side.
$4 > a$ Simplify.
$a < 4$ $4 > a$ is the same as $a < 4$.

The solution in set-builder notation is $\{a \mid a < 4\}$.
Number line graph:

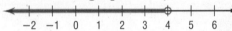

Exercises

Solve each inequality. Then check your solution, and graph it on a number line.

1. $t - 12 \geq 16$

   26 27 28 29 30 31 32 33 34

2. $n - 12 < 6$

   12 13 14 15 16 17 18 19 20

3. $6 \leq g - 3$

   7 8 9 10 11 12 13 14 15

4. $n - 8 < -13$

   −10 −9 −8 −7 −6 −5 −4 −3 −2

5. $-12 > -12 + y$

   −4 −3 −2 −1 0 1 2 3 4

6. $-6 > s - 8$

   −4 −3 −2 −1 0 1 2 3 4

Solve each inequality. Then check your solution.

7. $-3x \leq 8 - 4x$

8. $0.6n \geq 12 - 0.4n$

9. $-8k - 12 < -9k$

10. $-y - 10 > 15 - 2y$

11. $z - \dfrac{1}{3} \leq \dfrac{4}{3}$

12. $-2b > -4 - 3b$

Define a variable, write an inequality, and solve each problem. Then check your solution.

13. A number decreased by 4 is less than 14.

14. The difference of two numbers is more than 12, and one of the numbers is 3.

15. Forty is no greater than the difference of a number and 2.

Study Guide and Intervention   Glencoe Algebra 1

NAME _____ DATE _____ PERIOD _____

6-1 Study Guide and Intervention (continued)
Solving Inequalities by Addition and Subtraction

Solve Inequalities by Subtraction Subtraction can be used to solve inequalities. If any number is subtracted from each side of a true inequality, the resulting inequality is also true.

Subtraction Property of Inequalities	For all numbers a, b, and c, if $a > b$, then $a - c > b - c$, and if $a < b$, then $a - c < b - c$.

The property is also true when $>$ and $<$ are replaced with $\geq$ and $\leq$.

Example Solve $3a + 5 > 4 + 2a$. Then graph it on a number line.

$3a + 5 > 4 + 2a$ Original inequality
$3a + 5 - 2a > 4 + 2a - 2a$ Subtract 2a from each side.
$a + 5 > 4$ Simplify.
$a + 5 - 5 > 4 - 5$ Subtract 5 from each side.
$a > -1$ Simplify.

The solution is $\{a \mid a > -1\}$.

Number line graph:

Exercises

Solve each inequality. Then check your solution, and graph it on a number line.

1. $t + 12 \geq 8$ 2. $n + 12 > -12$ 3. $16 \leq h + 9$

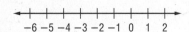   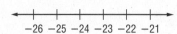   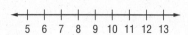

4. $y + 4 > -2$ 5. $3r + 6 > 4r$ 6. $\dfrac{3}{2}q - 5 \geq \dfrac{1}{2}q$

        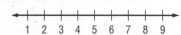   

Solve each inequality. Then check your solution.

7. $4p \geq 3p + 0.7$ 8. $r + \dfrac{1}{4} > \dfrac{3}{8}$ 9. $9k + 12 > 8k$

10. $-1.2 > 2.4 + y$ 11. $4y < 5y + 14$ 12. $3n + 17 < 4n$

Define a variable, write an inequality, and solve each problem. Then check your solution.

13. The sum of a number and 8 is less than 12.

14. The sum of two numbers is at most 6, and one of the number is -2.

15. The sum of a number and 6 is greater than or equal to -4.

NAME _____ DATE _____ PERIOD _____

6-2 Study Guide and Intervention

Solving Inequalities by Multiplication and Division

Solve Inequalities by Multiplication If each side of an inequality is multiplied by the same positive number, the resulting inequality is also true. However, if each side of an inequality is multiplied by the same negative number, the direction of the inequality must be reversed for the resulting inequality to be true.

Multiplication Property of Inequalities	For all numbers a, b, and c, with $c \neq 0$, 1. if c is positive and $a > b$, then $ac > bc$; if c is positive and $a < b$, then $ac < bc$; 2. if c is negative and $a > b$, then $ac < bc$; if c is negative and $a < b$, then $ac > bc$.

The property is also true when $>$ and $<$ are replaced with $\geq$ and $\leq$.

Example 1 Solve $-\frac{y}{8} \geq 12$.

$-\frac{y}{8} \geq 12$ Original equation

$(-8)\left(-\frac{y}{8}\right) \leq (-8)12$ Multiply each side by -8; change $\geq$ to $\leq$.

$y \leq -96$ Simplify.

The solution is $\{y \mid y \leq -96\}$.

Example 2 Solve $\frac{3}{4}k < 15$.

$\frac{3}{4}k < 15$ Original equation

$\left(\frac{4}{3}\right)\frac{3}{4}k < \left(\frac{4}{3}\right)15$ Multiply each side by $\frac{4}{3}$.

$k < 20$ Simplify.

The solution is $\{k \mid k < 20\}$.

Exercises

Solve each inequality. Then check your solution.

1. $\frac{y}{6} \leq 2$

2. $-\frac{n}{50} > 22$

3. $\frac{3}{5}h \geq -3$

4. $-\frac{p}{6} < -6$

5. $\frac{1}{4}n \geq 10$

6. $-\frac{2}{3}b < \frac{1}{3}$

7. $\frac{3m}{5} < -\frac{3}{20}$

8. $-2.51 \leq -\frac{2h}{4}$

9. $\frac{g}{5} \geq -2$

10. $-\frac{3}{4} > -\frac{9p}{5}$

11. $\frac{n}{10} \geq 5.4$

12. $\frac{2a}{7} \geq -6$

Define a variable, write an inequality, and solve each problem. Then check your solution.

13. Half of a number is at least 14.

14. The opposite of one-third a number is greater than 9.

15. One fifth of a number is at most 30.

Study Guide and Intervention   Glencoe Algebra 1

NAME _____ DATE _____ PERIOD _____

6-2 Study Guide and Intervention (continued)
Solving Inequalities by Multiplication and Division

Solve Inequalities by Division If each side of a true inequality is divided by the same positive number, the resulting inequality is also true. However, if each side of an inequality is divided by the same negative number, the direction of the inequality symbol must be reversed for the resulting inequality to be true.

Division Property of Inequalities	For all numbers a, b, and c with $c \neq 0$, 1. if c is positive and $a > b$, then $\frac{a}{c} > \frac{b}{c}$; if c is positive and $a < b$, then $\frac{a}{c} < \frac{b}{c}$; 2. if c is negative and $a > b$, then $\frac{a}{c} < \frac{b}{c}$; if c is negative and $a < b$, then $\frac{a}{c} > \frac{b}{c}$.

The property is also true when $>$ and $<$ are replaced with $\geq$ and $\leq$.

Example
Solve $-12y \geq 48$.

$-12y \geq 48$ Original inequality

$\frac{-12y}{-12} \leq \frac{48}{-12}$ Divide each side by -12 and change $\geq$ to $\leq$.

$y \leq -4$ Simplify.

The solution is $\{y \mid y \leq -4\}$.

Exercises

Solve each inequality. Then check your solution.

1. $25g \geq -100$
2. $-2x \geq 9$
3. $-5c > 2$
4. $-8m < -64$

5. $-6k < \frac{1}{5}$
6. $18 < -3b$
7. $30 < -3n$
8. $-0.24 < 0.6w$

9. $25 \geq -2m$
10. $-30 > -5p$
11. $-2n \geq 6.2$
12. $-35 < 0.05h$

13. $-40 > 10h$
14. $-\frac{2}{3}n \geq 6$
15. $-3 < \frac{p}{4}$

Define a variable, write an inequality, and solve each problem. Then check your solution.

16. Four times a number is no more than 108.

17. The opposite of three times a number is greater than 12.

18. Negative five times a number is at most 100.

6-3 Study Guide and Intervention

Solving Multi-Step Inequalities

Solve Multi-Step Inequalities To solve linear inequalities involving more than one operation, undo the operations in reverse of the order of operations, just as you would solve an equation with more than one operation.

Example 1 Solve $6x - 4 \leq 2x + 12$.

$6x - 4 \leq 2x + 12$	Original inequality
$6x - 4 - 2x \leq 2x + 12 - 2x$	Subtract 2x from each side.
$4x - 4 \leq 12$	Simplify.
$4x - 4 + 4 \leq 12 + 4$	Add 4 to each side.
$4x \leq 16$	Simplify.
$\dfrac{4x}{4} \leq \dfrac{16}{4}$	Divide each side by 4.
$x \leq 4$	Simplify.

The solution is $\{x \mid x \leq 4\}$.

Example 2 Solve $3a - 15 > 4 + 5a$.

$3a - 15 > 4 + 5a$	Original inequality
$3a - 15 - 5a > 4 + 5a - 5a$	Subtract 5a from each side.
$-2a - 15 > 4$	Simplify.
$-2a - 15 + 15 > 4 + 15$	Add 15 to each side.
$-2a > 19$	Simplify.
$\dfrac{-2a}{-2} < \dfrac{19}{-2}$	Divide each side by -2 and change $>$ to $<$.
$a < -9\dfrac{1}{2}$	Simplify.

The solution is $\left\{a \mid a < -9\dfrac{1}{2}\right\}$.

Exercises

Solve each inequality. Then check your solution.

1. $11y + 13 \geq -1$

2. $8n - 10 < 6 - 2n$

3. $\dfrac{q}{7} + 1 > -5$

4. $6n + 12 < 8 + 8n$

5. $-12 - d > -12 + 4d$

6. $5r - 6 > 8r - 18$

7. $\dfrac{-3x + 6}{2} \leq 12$

8. $7.3y - 14.4 > 4.9y$

9. $-8m - 3 < 18 - m$

10. $-4y - 10 > 19 - 2y$

11. $9n - 24n + 45 > 0$

12. $\dfrac{4x - 2}{5} \geq -4$

Define a variable, write an inequality, and solve each problem. Then check your solution.

13. Negative three times a number plus four is no more than the number minus eight.

14. One fourth of a number decreased by three is at least two.

15. The sum of twelve and a number is no greater than the sum of twice the number and -8.

NAME _____ DATE _____ PERIOD _____

6-3 Study Guide and Intervention (continued)
Solving Multi-Step Inequalities

Solve Inequalities Involving the Distributive Property When solving inequalities that contain grouping symbols, first use the Distributive Property to remove the grouping symbols. Then undo the operations in reverse of the order of operations, just as you would solve an equation with more than one operation.

Example 1 Solve $3a - 2(6a - 4) > 4 - (4a + 6)$.

$3a - 2(6a - 4) > 4 - (4a + 6)$	Original inequality
$3a - 12a + 8 > 4 - 4a - 6$	Distributive Property
$-9a + 8 > -2 - 4a$	Combine like terms.
$-9a + 8 + 4a > -2 - 4a + 4a$	Add 4a to each side.
$-5a + 8 > -2$	Combine like terms.
$-5a + 8 - 8 > -2 - 8$	Subtract 8 from each side.
$-5a > -10$	Simplify.
$a < 2$	Divide each side by -5 and change $>$ to $<$.

The solution in set-builder notation is $\{a \mid a < 2\}$.

Exercises

Solve each inequality. Then check your solution.

1. $2(t + 3) \geq 16$

2. $3(d - 2) - 2d > 16$

3. $4h - 8 < 2(h - 1)$

4. $6y + 10 > 8 - (y + 14)$

5. $4.6(x - 3.4) > 5.1x$

6. $-5x - (2x + 3) \geq 1$

7. $3(2y - 4) - 2(y + 1) > 10$

8. $8 - 2(b + 1) < 12 - 3b$

9. $-2(k - 1) > 8(1 + k)$

10. $0.3(y - 2) > 0.4(1 + y)$

11. $m + 17 \leq -(4m - 13)$

12. $3n + 8 \leq 2(n - 4) - 2(1 - n)$

13. $2(y - 2) > -4 + 2y$

14. $k - 17 \leq -(17 - k)$

15. $n - 4 \leq -3(2 + n)$

Define a variable, write an inequality, and solve each problem. Then check your solution.

16. Twice the sum of a number and 4 is less than 12.

17. Three times the sum of a number and six is greater than four times the number decreased by two.

18. Twice the difference of a number and four is less than the sum of the number and five.

Study Guide and Intervention 76 Glencoe Algebra 1

NAME _____ DATE _____ PERIOD _____

6-4 Study Guide and Intervention

Solving Compound Inequalities

Inequalities Containing *and* A compound inequality containing *and* is true only if both inequalities are true. The graph of a compound inequality containing *and* is the **intersection** of the graphs of the two inequalities. Every solution of the compound inequality must be a solution of both inequalities.

Example 1 Graph the solution set of $x < 2$ and $x \geq -1$.

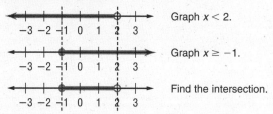

The solution set is $\{x \mid -1 \leq x < 2\}$.

Example 2 Solve $-1 < x + 2 < 3$ using *and*. Then graph the solution set.

$-1 < x + 2$ and $x + 2 < 3$
$-1 - 2 < x + 2 - 2$ $\quad\quad x + 2 - 2 < 3 - 2$
$-3 < x$ $\quad\quad\quad\quad\quad\quad x < 1$

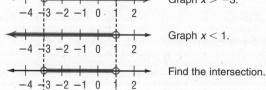

The solution set is $\{x \mid -3 < x < 1\}$.

Exercises

Graph the solution set of each compound inequality.

1. $b > -1$ and $b \leq 3$

2. $2 \geq q \geq -5$

3. $x > -3$ and $x \leq 4$

4. $-2 \leq p < 4$

5. $-3 < d$ and $d < 2$

6. $-1 \leq p \leq 3$

Solve each compound inequality. Then graph the solution set.

7. $4 < w + 3 \leq 5$

8. $-3 \leq p - 5 < 2$

9. $-4 < x + 2 \leq -2$

10. $y - 1 < 2$ and $y + 2 \geq 1$

11. $n - 2 > -3$ and $n + 4 < 6$

12. $d - 3 < 6d + 12 < 2d + 32$

Study Guide and Intervention

Glencoe Algebra 1

6-4 Study Guide and Intervention (continued)
Solving Compound Inequalities

Inequalities Containing *or* A compound inequality containing *or* is true if one or both of the inequalities are true. The graph of a compound inequality containing *or* is the **union** of the graphs of the two inequalities. The union can be found by graphing both inequalities on the same number line. A solution of the compound inequality is a solution of either inequality, not necessarily both.

Example Solve $2a + 1 < 11$ or $a > 3a + 2$.

$$2a + 1 < 11 \quad\quad\text{or}\quad\quad a > 3a + 2$$
$$2a + 1 - 1 < 11 - 1 \quad\quad a - 3a > 3a - 3a + 2$$
$$2a < 10 \quad\quad -2a > 2$$
$$\frac{2a}{2} < \frac{10}{2} \quad\quad \frac{-2a}{-2} < \frac{2}{-2}$$
$$a < 5 \quad\quad a < -1$$

Graph $a < 5$.

Graph $a < -1$.

Find the union.

The solution set is $\{a \mid a < 5\}$.

Exercises

Graph the solution set of each compound inequality.

1. $b > 2$ or $b \leq -3$

2. $3 \geq q$ or $q \leq 1$

3. $y \leq -4$ or $y > 0$

4. $4 \leq p$ or $p < 8$

5. $-3 < d$ or $d < 2$

6. $-2 \leq x$ or $3 \leq x$

Solve each compound inequality. Then graph the solution set.

7. $3 < 3w$ or $3w \geq 9$

8. $-3p + 1 \leq -11$ or $p < 2$

9. $2x + 4 \leq 6$ or $x \geq 2x - 4$

10. $2y + 2 < 12$ or $y - 3 \geq 2y$

11. $\frac{1}{2}n > -2$ or $2n - 2 < 6 + n$

12. $3a + 2 \geq 5$ or $7 + 3a < 2a + 6$

NAME _____ DATE _____ PERIOD _____

6-5 Study Guide and Intervention
Solving Open Sentences Involving Absolute Value

Absolute Value Equations When solving equations that involve absolute value, there are two cases to consider.

Case 1: The value inside the absolute value symbols is positive.
Case 2: The value inside the absolute value symbols is negative.

Example 1 Solve $|x + 4| = 1$. Then graph the solution set.

Write $|x + 4| = 1$ as $x + 4 = 1$ or $x + 4 = -1$.

$x + 4 = 1$ or $x + 4 = -1$
$x + 4 - 4 = 1 - 4$ $x + 4 = -1$
$x = -3$ $x + 4 - 4 = -1 - 4$
     $x = -5$

The solution set is $\{-5, -3\}$.
The graph is shown below.

−8 −7 −6 −5 −4 −3 −2 −1 0

Example 2 Write an inequality involving absolute value for the graph.

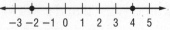

Find the point that is the same distance from -2 as it is from 4.

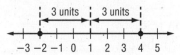

The distance from 1 to -2 is 3 units. The distance from 1 to 4 is 3 units.
So, $|x - 1| = 3$.

Exercises

Solve each open sentence. Then graph the solution set.

1. $|y| = 3$

   −4 −3 −2 −1 0 1 2 3 4

2. $|x - 4| = 4$

   0 1 2 3 4 5 6 7 8

3. $|y + 3| = 2$

   −8 −7 −6 −5 −4 −3 −2 −1 0

4. $|b + 2| = 3$

   −6 −5 −4 −3 −2 −1 0 1 2

5. $|w - 2| = 5$

   −8 −6 −4 −2 0 2 4 6 8

6. $|t + 2| = 4$

   −8 −6 −4 −2 0 2 4 6 8

7. $|2x| = 8$

   −4 −3 −2 −1 0 1 2 3 4

8. $|5y - 2| = 7$

   −4 −3 −2 −1 0 1 2 3 4

9. $|p - 0.2| = 0.5$

   −0.8 −0.4   0   0.4   0.8

10. $|d - 100| = 50$

    50 100 150 200

11. $|2x - 1| = 11$

    −6 −4 −2 0 2 4 6 8 10

12. $\left|3x + \dfrac{1}{2}\right| = 6$

    −3 −2 −1 0 1 2 3 4 5

For each graph, write an open sentence involving absolute value.

13.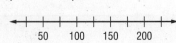
−8 −6 −4 −2 0 2 4 6 8

14.
−4 −3 −2 −1 0 1 2 3 4

15.
−7 −6 −5 −4 −3 −2 −1 0 1

NAME _____ DATE _____ PERIOD _____

6-5 Study Guide and Intervention (continued)

Solving Open Sentences Involving Absolute Value

Absolute Value Equations When solving equations that involve absolute value, there are two cases to consider.

Case 1: The value inside the absolute value symbols is positive.
Case 2: The value inside the absolute value symbols is negative.

Example 1 Solve $|x + 1| = 2$. Then graph the solution set.

Write $|x + 1| = 2$ as $x + 1 = 2$ and $x + 1 = -2$.

$x + 1 = 2$ or $x + 1 = -2$
$x + 1 - 1 = 2 - 1$ $x + 1 - 1 = -2 - 1$
$x = 1$ $x = -3$

The solution set is $\{-3, 1\}$.
The graph is shown below.

$\xleftarrow{\quad}\underset{-5\;-4\;-3\;-2\;-1\;\;0\;\;1\;\;2\;\;3}{\bullet\;\bullet}\xrightarrow{\quad}$

Example 2 Write an equation involving absolute value for the graph.

$\xleftarrow{\quad}\underset{-3\;-2\;-1\;\;0\;\;1\;\;2\;\;3\;\;4\;\;5}{\;\;\;\;\;\;\;\;\;\;\bullet\;\;\;\;\;\;\;\;\;\;\;\;\;\;\bullet}\xrightarrow{\quad}$

Find the point that is the same distance from -1 as it is from 3.
The distance from 1 to -1 is 2 units.
The distance from 1 to 3 is 2 units.
So, $|x - 1| = 2$

Exercises

Solve each open sentence. Then graph the solutions set.

1. $|w - 3| = 4$

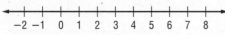

2. $|2k - 3| = 1$

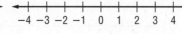

3. $|x - 3.2| = 0.8$

For each graph, write an open sentence involving absolute value.

4. 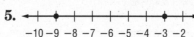   5. 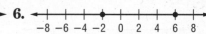   6.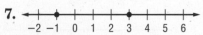

7. 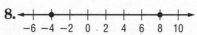   8. 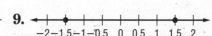   9.

10. 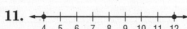   11. (graph)   12. (graph)

13. (graph)   14. (graph)   15. (graph)

NAME _____ DATE _____ PERIOD _____

6-6 Study Guide and Intervention

Solving Inequalities Involving Absolute Value

Absolute Value Inequalities When solving inequalities that involve absolute value, there are two cases to consider for inequalities involving $<$ (or $\leq$) and two cases to consider for inequalities involving $>$ (or $\geq$).

If $|x| < n$, then $x > -n$ and $x < n$.
If $|x| > n$, then $x > n$ or $x < -n$.

Remember that inequalities with *and* are related to intersections, while inequalities with *or* are related to unions.

Example Solve $|3a + 4| < 10$. Then graph the solution set.

Write $|3a + 4| < 10$ as $3a + 4 < 10$ and $3a + 4 > -10$.

$3a + 4 < 10$ and $3a + 4 > -10$
$3a + 4 - 4 < 10 - 4$ $\qquad$ $3a + 4 - 4 > -10 - 4$
$3a < 6$ $\qquad$ $3a > -14$
$\dfrac{3a}{3} < \dfrac{6}{3}$ $\qquad$ $\dfrac{3a}{3} > \dfrac{-14}{3}$
$a < 2$ $\qquad$ $a > -4\dfrac{2}{3}$

The solution set is $\left\{a \mid -4\dfrac{2}{3} < a < 2\right\}$.

Now graph the solution set.

$\leftarrow\!\!\circ\!\!+\!\!+\!\!+\!\!+\!\!+\!\!+\!\!\circ\!\!+\!\!\rightarrow$
$-5\ -4\ -3\ -2\ -1\ \ 0\ \ 1\ \ 2\ \ 3$

Exercises

Solve each open sentence. Then graph the solution set.

1. $|c - 2| > 6$

$\leftarrow\!\!+\!\!+\!\!+\!\!+\!\!+\!\!+\!\!+\!\!+\!\!\rightarrow$
$-4\ -2\ \ 0\ \ 2\ \ 4\ \ 6\ \ 8\ \ 10$

2. $|x - 3| < 0$

$\leftarrow\!\!+\!\!+\!\!+\!\!+\!\!+\!\!+\!\!+\!\!+\!\!+\!\!\rightarrow$
$-4\ -3\ -2\ -1\ \ 0\ \ 1\ \ 2\ \ 3\ \ 4$

3. $|3f + 10| \leq 4$

$\leftarrow\!\!+\!\!+\!\!+\!\!+\!\!+\!\!+\!\!+\!\!\rightarrow$
$-5\ -4\ -3\ -2\ -1\ \ 0\ \ 1\ \ 2$

4. $|x| \leq 2$

$\leftarrow\!\!+\!\!+\!\!+\!\!+\!\!+\!\!+\!\!+\!\!+\!\!\rightarrow$
$-4\ -3\ -2\ -1\ \ 0\ \ 1\ \ 2\ \ 3\ \ 4$

5. $|x| \geq 3$

$\leftarrow\!\!+\!\!+\!\!+\!\!+\!\!+\!\!+\!\!+\!\!\rightarrow$
$-3\ -2\ -1\ \ 0\ \ 1\ \ 2\ \ 3\ \ 4$

6. $|2x + 1| \geq -2$

$\leftarrow\!\!+\!\!+\!\!+\!\!+\!\!+\!\!+\!\!+\!\!+\!\!\rightarrow$
$-3\ -2\ -1\ \ 0\ \ 1\ \ 2\ \ 3\ \ 4$

7. $|2d - 1| \leq 4$

8. $|3 - (x - 1)| \leq 8$

$\leftarrow\!\!+\!\!+\!\!+\!\!+\!\!+\!\!+\!\!+\!\!\rightarrow$
$-2\ \ 0\ \ 2\ \ 4\ \ 6\ \ 8\ \ 10\ \ 12$

9. $|3r + 2| < -5$

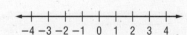

For each graph, write an open sentence involving absolute value.

10. $\leftarrow\!\!+\!\!+\!\!\circ\!\!+\!\!\circ\!\!+\!\!+\!\!\rightarrow$
$\ \ -4\ -3\ -2\ -1\ \ 0\ \ 1\ \ 2\ \ 3\ \ 4$

11. $\leftarrow\!\!+\!\!+\!\!\circ\!\!+\!\!\circ\!\!+\!\!\rightarrow$
$\ \ -4\ -3\ -2\ -1\ \ 0\ \ 1\ \ 2\ \ 3\ \ 4$

12.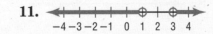

NAME _____ DATE _____ PERIOD _____

6-6 Study Guide and Intervention (continued)
Solving Inequalities Involving Absolute Value

Absolute Value Inequalities When solving inequalities that involve absolute value, there are two cases to consider for inequalities involving $<$ (or $\leq$) and two cases to consider for inequalities involving $>$ (or $\geq$).

> If $|x| < n$, then $x > -n$ and $x < n$.
> If $|x| > n$, then $x > n$ or $x > -n$.

Remember that inequalities with *and* are related to intersections, while inequalities with *or* are related to unions.

Example 1 Solve $|2x + 3| > 5$. Then graph the solution set.

Write $|2x + 3| > 5$ as $2x + 3 > 5$ and $2x + 3 < -5$.

$2x + 3 > 5$ or $2x + 3 < -5$
$2x + 3 - 3 > 5 - 3$ $2x + 3 - 3 < -5 - 3$
$\quad\quad 2x > 2$ $2x < -8$
$\quad\quad \dfrac{2x}{2} > \dfrac{2}{2}$ $\dfrac{2x}{2} < \dfrac{-8}{2}$
$\quad\quad\quad x > 1$ $x < -4$

The solution set is $x < -4$ or $x > 1$.
Now graph the solution set.

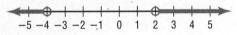

Example 2 Write an inequality involving absolute value from the graph.

Find the point that is the same distance form -4 as it is from 2.
The distance from -4 to -1 is 3 units.
The distance from 2 to -1 is 3 units.
The solution set is $\{x \mid -4 < x < 2\}$.
So, $|x + 2| < 3$

Exercises

Solve each open sentence. Then graph the solution set.

1. $|b - 6| > 3$

2. $|f - 4| < 1$

3. $|2m + 5| \geq 7$

For each graph, write an open sentence involving absolute value.

4.

5.

6.

7.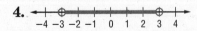

8.

9.

Study Guide and Intervention   Glencoe Algebra 1

NAME _____ DATE _____ PERIOD _____

6-7 Study Guide and Intervention

Graphing Inequalities in Two Variables

Graph Linear Inequalities The solution set of an inequality that involves two variables is graphed by graphing a related linear equation that forms a boundary of a **half-plane**. The graph of the ordered pairs that make up the solution set of the inequality fill a region of the coordinate plane on one side of the half-plane.

Example Graph $y \leq -3x - 2$.

Graph $y = -3x - 2$.
Since $y \leq -3x - 2$ is the same as $y < -3x - 2$ and $y = -3x - 2$, the boundary is included in the solution set and the graph should be drawn as a solid line.
Select a point in each half plane and test it. Choose (0, 0) and (−2, −2).

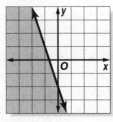

$y \leq -3x - 2$ $y \leq -3x - 2$
$0 \leq -3(0) - 2$ $-2 \leq -3(-2) - 2$
$0 \leq -2$ is false. $-2 \leq 6 - 2$
                          $-2 \leq 4$ is true.

The half-plane that contains (−2, −2) contains the solution. Shade that half-plane.

Exercises

Graph each inequality.

1. $y < 4$ 2. $x \geq 1$ 3. $3x \leq y$

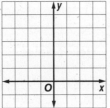 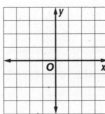 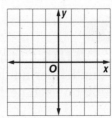

4. $-x > y$ 5. $x - y \geq 1$ 6. $2x - 3y \leq 6$

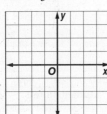 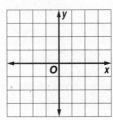 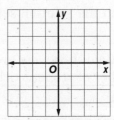

7. $y < -\frac{1}{2}x - 3$ 8. $4x - 3y < 6$ 9. $3x + 6y \geq 12$

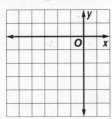 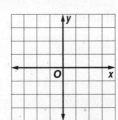 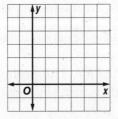

Study Guide and Intervention   Glencoe Algebra 1

NAME _____ DATE _____ PERIOD _____

6-7 Study Guide and Intervention (continued)

Graphing Inequalities in Two Variables

Solve Real-World Problems When solving real-life inequalities, the domain and range of the inequality are often restricted to nonnegative numbers or to whole numbers.

Example BANKING A bank offers 4.5% annual interest on regular savings accounts and 6% annual interest on certificates of deposit (CD). If Marjean wants to earn at least $300 interest per year, how much money should she deposit in each type of account?

Let x = the amount deposited in a regular savings account.
Let y = the amount deposited in a CD.
Then $0.045x + 0.06y \geq 300$ is an open sentence representing this situation.

Solve for y in terms of x.

$0.045x + 0.06y \geq 300$ Original inequality
$0.06y \geq -0.045x + 300$ Subtract 0.045x from each side.
$y \geq -0.75x + 5000$ Divide each side by 0.06.

Graph $y \geq -0.75x + 5000$ and test the point (0, 0). Since $0 \geq -0.75(0) + 5000$ is false, shade the half-plane that does not contain (0, 0).

One solution is (4000, 2000). This represents $4000 deposited at 4.5% and $2000 deposited at 6%.

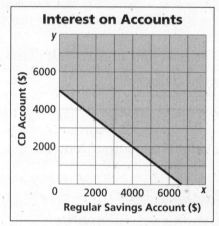

Interest on Accounts

Exercises

1. **SOCIAL EVENTS** Tickets for the school play cost $5 per student and $7 per adult. The school wants to earn at least $5400 on each performance.

   a. Write an inequality that represents this situation.

   b. Graph the solution set.

   c. If 500 adult tickets are sold, what is the minimum number of student tickets that must be sold?

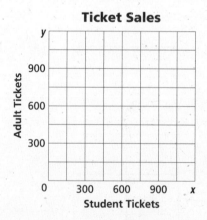

Ticket Sales

2. **MANUFACTURING** An auto parts company can produce 525 four-cylinder engines or 270 V-6 engines per day. It wants to produce up to 300,000 engines per year.

   a. Write an inequality that represents this situation.

   b. Are there restrictions on the domain or range?

3. **GEOMETRY** The perimeter of a rectangular lot is less than 800 feet. Write an inequality that represents the amount of fencing that will enclose the lot.

NAME _____ DATE _____ PERIOD _____

6-8 Study Guide and Intervention

Graphing Systems of Inequalities

Systems of Inequalities The solution of a **system of inequalities** is the set of all ordered pairs that satisfy both inequalities. If you graph the inequalities in the same coordinate plane, the solution is the region where the graphs overlap.

Example 1 Solve the system of inequalities by graphing.
$y > x + 2$
$y \leq -2x - 1$

The solution includes the ordered pairs in the intersection of the graphs. This region is shaded at the right. The graphs of $y = x + 2$ and $y = -2x - 1$ are boundaries of this region. The graph of $y = x + 2$ is dashed and is not included in the graph of $y > x + 2$.

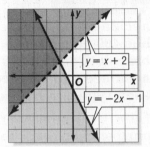

Example 2 Solve the system of inequalities by graphing.
$x + y > 4$
$x + y < -1$

The graphs of $x + y = 4$ and $x + y = -1$ are parallel. Because the two regions have no points in common, the system of inequalities has no solution.

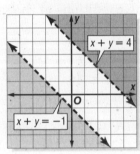

Exercises

Solve each system of inequalities by graphing.

1. $y > -1$
   $x < 0$

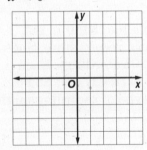

2. $y > -2x + 2$
   $y \leq x + 1$

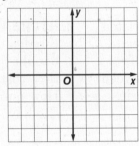

3. $y < x + 1$
   $3x + 4y \geq 12$

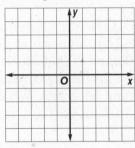

4. $2x + y \geq 1$
   $x - y \geq -2$

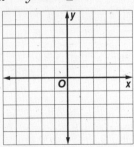

5. $y \leq 2x + 3$
   $y \geq -1 + 2x$

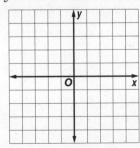

6. $5x - 2y < 6$
   $y > -x + 1$

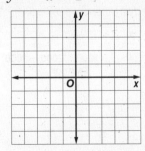

Study Guide and Intervention   Glencoe Algebra 1

NAME _____ DATE _____ PERIOD _____

6-8 Study Guide and Intervention (continued)

Graphing Systems of Inequalities

Real-World Problems In real-world problems, sometimes only whole numbers make sense for the solution, and often only positive values of x and y make sense.

Example **BUSINESS** AAA Gem Company produces necklaces and bracelets. In a 40-hour week, the company has 400 gems to use. A necklace requires 40 gems and a bracelet requires 10 gems. It takes 2 hours to produce a necklace and a bracelet requires one hour. How many of each type can be produced in a week?

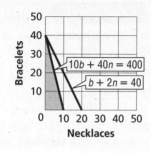

Let n = the number of necklaces that will be produced and b = the number of bracelets that will be produced. Neither n or b can be a negative number, so the following system of inequalities represents the conditions of the problems.

$n \geq 0$
$b \geq 0$
$b + 2n \leq 40$
$10b + 40n \leq 400$

The solution is the set ordered pairs in the intersection of the graphs. This region is shaded at the right. Only whole-number solutions, such as (5, 20) make sense in this problem.

Exercises

For each exercise, graph the solution set. List three possible solutions to the problem.

1. **HEALTH** Mr. Flowers is on a restricted diet that allows him to have between 1600 and 2000 Calories per day. His daily fat intake is restricted to between 45 and 55 grams. What daily Calorie and fat intakes are acceptable?

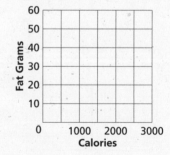

2. **RECREATION** Maria had $150 in gift certificates to use at a record store. She bought fewer than 20 recordings. Each tape cost $5.95 and each CD cost $8.95. How many of each type of recording might she have bought?

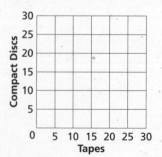

NAME _____ DATE _____ PERIOD _____

7-1 Study Guide and Intervention

Multiplying Monomials

Multiply Monomials A **monomial** is a number, a variable, or a product of a number and one or more variables. An expression of the form x^n is called a **power** and represents the product you obtain when x is used as a factor n times. To multiply two powers that have the same base, add the exponents.

Product of Powers	For any number a and all integers m and n, $a^m \cdot a^n = a^{m+n}$.

Example 1 Simplify $(3x^6)(5x^2)$.

$(3x^6)(5x^2) = (3)(5)(x^6 \cdot x^2)$ Group the coefficients and the variables

$= (3 \cdot 5)(x^{6+2})$ Product of Powers

$= 15x^8$ Simplify.

The product is $15x^8$.

Example 2 Simplify $(-4a^3b)(3a^2b^5)$.

$(-4a^3b)(3a^2b^5) = (-4)(3)(a^3 \cdot a^2)(b \cdot b^5)$

$= -12(a^{3+2})(b^{1+5})$

$= -12a^5b^6$

The product is $-12a^5b^6$.

Exercises

Simplify.

1. $y(y^5)$

2. $n^2 \cdot n^7$

3. $(-7x^2)(x^4)$

4. $x(x^2)(x^4)$

5. $m \cdot m^5$

6. $(-x^3)(-x^4)$

7. $(2a^2)(8a)$

8. $(rs)(rs^3)(s^2)$

9. $(x^2y)(4xy^3)$

10. $\frac{1}{3}(2a^3b)(6b^3)$

11. $(-4x^3)(-5x^7)$

12. $(-3j^2k^4)(2jk^6)$

13. $(5a^2bc^3)\left(\frac{1}{5}abc^4\right)$

14. $(-5xy)(4x^2)(y^4)$

15. $(10x^3yz^2)(-2xy^5z)$

Study Guide and Intervention

Glencoe Algebra 1

NAME _____ DATE _____ PERIOD _____

7-1 Study Guide and Intervention *(continued)*

Multiplying Monomials

Powers of Monomials An expression of the form $(x^m)^n$ is called a **power of a power** and represents the product you obtain when x^m is used as a factor n times. To find the power of a power, multiply exponents.

Power of a Power	For any number a and all integers m and n, $(a^m)^n = a^{mn}$.
Power of a Product	For any number a and all integers m and n, $(ab)^m = a^m b^m$.

Example Simplify $(-2ab^2)^3(a^2)^4$.

$(-2ab^2)^3(a^2)^4 = (-2ab^2)^3(a^8)$ Power of a Power
$= (-2)^3(a^3)(b^2)^3(a^8)$ Power of a Product
$= (-2)^3(a^3)(a^8)(b^2)^3$ Group the coefficients and the variables
$= (-2)^3(a^{11})(b^2)^3$ Product of Powers
$= -8a^{11}b^6$ Power of a Power

The product is $-8a^{11}b^6$.

Exercises

Simplify.

1. $(y^5)^2$

2. $(n^7)^4$

3. $(x^2)^5(x^3)$

4. $-3(ab^4)^3$

5. $(-3ab^4)^3$

6. $(4x^2b)^3$

7. $(4a^2)^2(b^3)$

8. $(4x)^2(b^3)$

9. $(x^2y^4)^5$

10. $(2a^3b^2)(b^3)^2$

11. $(-4xy)^3(-2x^2)^3$

12. $(-3j^2k^3)^2(2j^2k)^3$

13. $(25a^2b)^3\left(\dfrac{1}{5}abc\right)^2$

14. $(2xy)^2(-3x^2)(4y^4)$

15. $(2x^3y^2z^2)^3(x^2z)^4$

16. $(-2n^6y^5)(-6n^3y^2)(ny)^3$

17. $(-3a^3n^4)(-3a^3n)^4$

18. $-3(2x)^4(4x^5y)^2$

Study Guide and Intervention   Glencoe Algebra 1

NAME _____ DATE _____ PERIOD _____

7-2 Study Guide and Intervention
Dividing Monomials

Quotients of Monomials To divide two powers with the same base, subtract the exponents.

Quotient of Powers	For all integers m and n and any nonzero number a, $\dfrac{a^m}{a^n} = a^{m-n}$.
Power of a Quotient	For any integer m and any real numbers a and b, $b \neq 0$, $\left(\dfrac{a}{b}\right)^m = \dfrac{a^m}{b^m}$.

Example 1 Simplify $\dfrac{a^4 b^7}{ab^2}$. Assume neither a nor b is equal to zero.

$\dfrac{a^4 b^7}{ab^2} = \left(\dfrac{a^4}{a}\right)\left(\dfrac{b^7}{b^2}\right)$ Group powers with the same base.

$= (a^{4-1})(b^{7-2})$ Quotient of Powers

$= a^3 b^5$ Simplify.

The quotient is $a^3 b^5$.

Example 2 Simplify $\left(\dfrac{2a^3 b^5}{3b^2}\right)^3$. Assume that b is not equal to zero.

$\left(\dfrac{2a^3 b^5}{3b^2}\right)^3 = \dfrac{(2a^3 b^5)^3}{(3b^2)^3}$ Power of a Quotient

$= \dfrac{2^3 (a^3)^3 (b^5)^3}{(3)^3 (b^2)^3}$ Power of a Product

$= \dfrac{8a^9 b^{15}}{27 b^6}$ Power of a Power

$= \dfrac{8a^9 b^9}{27}$ Quotient of Powers

The quotient is $\dfrac{8a^9 b^9}{27}$.

Exercises

Simplify. Assume that no denominator is equal to zero.

1. $\dfrac{5^5}{5^2}$

2. $\dfrac{m^6}{m^4}$

3. $\dfrac{p^5 n^4}{p^2 n}$

4. $\dfrac{a^2}{a}$

5. $\dfrac{x^5 y^3}{x^5 y^2}$

6. $\dfrac{-2y^7}{14 y^5}$

7. $\dfrac{xy^6}{y^4 x}$

8. $\left(\dfrac{2a^2 b}{a}\right)^3$

9. $\left(\dfrac{4p^4 q^4}{3p^2 q^2}\right)^3$

10. $\left(\dfrac{2v^5 w^3}{v^4 w^3}\right)^4$

11. $\left(\dfrac{3r^6 s^3}{2r^5 s}\right)^4$

12. $\dfrac{r^7 s^7 t^2}{s^3 r^3 t^2}$

Study Guide and Intervention Glencoe Algebra 1

NAME _____ DATE _____ PERIOD _____

7-2 Study Guide and Intervention (continued)
Dividing Monomials

Negative Exponents Any nonzero number raised to the zero power is 1; for example, $(-0.5)^0 = 1$. Any nonzero number raised to a negative power is equal to the reciprocal of the number raised to the opposite power; for example, $6^{-3} = \dfrac{1}{6^3}$. These definitions can be used to simplify expressions that have negative exponents.

Zero Exponent	For any nonzero number a, $a^0 = 1$.
Negative Exponent Property	For any nonzero number a and any integer n, $a^{-n} = \dfrac{1}{a^n}$ and $\dfrac{1}{a^{-n}} = a^n$.

The simplified form of an expression containing negative exponents must contain only positive exponents.

Example Simplify $\dfrac{4a^{-3}b^6}{16a^2b^6c^{-5}}$. Assume that the denominator is not equal to zero.

$\dfrac{4a^{-3}b^6}{16a^2b^6c^{-5}} = \left(\dfrac{4}{16}\right)\left(\dfrac{a^{-3}}{a^2}\right)\left(\dfrac{b^6}{b^6}\right)\left(\dfrac{1}{c^{-5}}\right)$ Group powers with the same base.

$= \dfrac{1}{4}(a^{-3-2})(b^{6-6})(c^5)$ Quotient of Powers and Negative Exponent Properties

$= \dfrac{1}{4}a^{-5}b^0c^5$ Simplify.

$= \dfrac{1}{4}\left(\dfrac{1}{a^5}\right)(1)c^5$ Negative Exponent and Zero Exponent Properties

$= \dfrac{c^5}{4a^5}$ Simplify.

The solution is $\dfrac{c^5}{4a^5}$.

Exercises

Simplify. Assume that no denominator is equal to zero.

1. $\dfrac{2^2}{2^{-3}}$

2. $\dfrac{m}{m^{-4}}$

3. $\dfrac{p^{-8}}{p^3}$

4. $\dfrac{b^{-4}}{b^{-5}}$

5. $\dfrac{(-x^{-1}y)^0}{4w^{-1}y^2}$

6. $\dfrac{(a^2b^3)^2}{(ab)^{-2}}$

7. $\dfrac{x^4y^0}{x^{-2}}$

8. $\dfrac{(6a^{-1}b)^2}{(b^2)^4}$

9. $\dfrac{(3st)^2u^{-4}}{s^{-1}t^2u^7}$

10. $\dfrac{s^{-3}t^{-5}}{(s^2t^3)^{-1}}$

11. $\left(\dfrac{4m^2n^2}{8m^{-1}\ell}\right)^0$

12. $\dfrac{(-2mn^2)^{-3}}{4m^{-6}n^4}$

NAME _____ DATE _____ PERIOD _____

7-3 Study Guide and Intervention

Polynomials

Degree of a Polynomial A **polynomial** is a monomial or a sum of monomials. A **binomial** is the sum of two monomials, and a **trinomial** is the sum of three monomials. Polynomials with more than three terms have no special name. The **degree** of a monomial is the sum of the exponents of all its variables. The **degree of the polynomial** is the same as the degree of the monomial term with the highest degree.

Example State whether each expression is a polynomial. If the expression is a polynomial, identify it as a *monomial, binomial,* or *trinomial*. Then give the degree of the polynomial.

Expression	Polynomial?	Monomial, Binomial, or Trinomial?	Degree of the Polynomial
$3x - 7xyz$	Yes. $3x - 7xyz = 3x + (-7xyz)$, which is the sum of two monomials	binomial	3
-25	Yes. -25 is a real number.	monomial	0
$7n^3 + 3n^{-4}$	No. $3n^{-4} = \dfrac{3}{n^4}$, which is not a monomial	none of these	—
$9x^3 + 4x + x + 4 + 2x$	Yes. The expression simplifies to $9x^3 + 7x + 4$, which is the sum of three monomials	trinomial	3

Exercises

State whether each expression is a polynomial. If the expression is a polynomial, identify it as a *monomial, binomial,* or *trinomial*.

1. 36

2. $\dfrac{3}{q^2} + 5$

3. $7x - x + 5$

4. $8g^2h - 7gh + 2$

5. $\dfrac{1}{4y^2} + 5y - 8$

6. $6x + x^2$

Find the degree of each polynomial.

7. $4x^2y^3z$

8. $-2abc$

9. $15m$

10. $s + 5t$

11. 22

12. $18x^2 + 4yz - 10y$

13. $x^4 - 6x^2 - 2x^3 - 10$

14. $2x^3y^2 - 4xy^3$

15. $-2r^8s^4 + 7r^2s - 4r^7s^6$

16. $9x^2 + yz^8$

17. $8b + bc^5$

18. $4x^4y - 8zx^2 + 2x^5$

19. $4x^2 - 1$

20. $9abc + bc - d^5$

21. $h^3m + 6h^4m^2 - 7$

NAME _____ DATE _____ PERIOD _____

7-3 Study Guide and Intervention (continued)
Polynomials

Write Polynomials in Order The terms of a polynomial are usually arranged so that the powers of one variable are in **ascending** (increasing) order or **descending** (decreasing) order.

Example 1 Arrange the terms of each polynomial so that the powers of x are in ascending order.

a. $x^4 - x^2 + 5x^3$
   $-x^2 + 5x^3 + x^4$

b. $8x^3y - y^2 + 6x^2y + xy^2$
   $-y^2 + xy^2 + 6x^2y + 8x^3y$

Example 2 Arrange the terms of each polynomial so that the powers of x are in descending order.

a. $x^4 + 4x^5 - x^2$
   $4x^5 + x^4 - x^2$

b. $-6xy + y^3 - x^2y^2 + x^4y^2$
   $x^4y^2 - x^2y^2 - 6xy + y^3$

Exercises

Arrange the terms of each polynomial so that the powers of x are in ascending order.

1. $5x + x^2 + 6$

2. $6x + 9 - 4x^2$

3. $4xy + 2y + 6x^2$

4. $6y^2x - 6x^2y + 2$

5. $x^4 + x^3 + x^2$

6. $2x^3 - x + 3x^7$

7. $-5cx + 10c^2x^3 + 15cx^2$

8. $-4nx - 5n^3x^3 + 5$

9. $4xy + 2y + 5x^2$

Arrange the terms of each polynomial so that the powers of x are in descending order.

10. $2x + x^2 - 5$

11. $20x - 10x^2 + 5x^3$

12. $x^2 + 4yx - 10x^5$

13. $9bx + 3bx^2 - 6x^3$

14. $x^3 + x^5 - x^2$

15. $ax^2 + 8a^2x^5 - 4$

16. $3x^3y - 4xy^2 - x^4y^2 + y^5$

17. $x^4 + 4x^3 - 7x^5 + 1$

18. $-3x^6 - x^5 + 2x^8$

19. $-15cx^2 + 8c^2x^5 + cx$

20. $24x^2y - 12x^3y^2 + 6x^4$

21. $-15x^3 + 10x^4y^2 + 7xy^2$

NAME _____ DATE _____ PERIOD _____

7-4 Study Guide and Intervention

Adding and Subtracting Polynomials

Add Polynomials To add polynomials, you can group like terms horizontally or write them in column form, aligning like terms vertically. **Like terms** are monomial terms that are either identical or differ only in their coefficients, such as $3p$ and $-5p$ or $2x^2y$ and $8x^2y$.

Example 1 Find $(2x^2 + x - 8) + (3x - 4x^2 + 2)$.

Horizontal Method
Group like terms.
$(2x^2 + x - 8) + (3x - 4x^2 + 2)$
$= [(2x^2 + (-4x^2)] + (x + 3x) + [(-8) + 2]$
$= -2x^2 + 4x - 6.$
The sum is $-2x^2 + 4x - 6$.

Example 2 Find $(3x^2 + 5xy) + (xy + 2x^2)$.

Vertical Method
Align like terms in columns and add.
$3x^2 + 5xy$
$(+)\ 2x^2 + xy$ Put the terms in descending order.
$5x^2 + 6xy$
The sum is $5x^2 + 6xy$.

Exercises

Find each sum.

1. $(4a - 5) + (3a + 6)$

2. $(6x + 9) + (4x^2 - 7)$

3. $(6xy + 2y + 6x) + (4xy - x)$

4. $(x^2 + y^2) + (-x^2 + y^2)$

5. $(3p^2 - 2p + 3) + (p^2 - 7p + 7)$

6. $(2x^2 + 5xy + 4y^2) + (-xy - 6x^2 + 2y^2)$

7. $(5p + 2q) + (2p^2 - 8q + 1)$

8. $(4x^2 - x + 4) + (5x + 2x^2 + 2)$

9. $(6x^2 + 3x) + (x^2 - 4x - 3)$

10. $(x^2 + 2xy + y^2) + (x^2 - xy - 2y^2)$

11. $(2a - 4b - c) + (-2a - b - 4c)$

12. $(6xy^2 + 4xy) + (2xy - 10xy^2 + y^2)$

13. $(2p - 5q) + (3p + 6q) + (p - q)$

14. $(2x^2 - 6) + (5x^2 + 2) + (-x^2 - 7)$

15. $(3z^2 + 5z) + (z^2 + 2z) + (z - 4)$

16. $(8x^2 + 4x + 3y^2 + y) + (6x^2 - x + 4y)$

7-4 Study Guide and Intervention (continued)
Adding and Subtracting Polynomials

Subtract Polynomials You can subtract a polynomial by adding its additive inverse. To find the additive inverse of a polynomial, replace each term with its additive inverse or opposite.

Example Find $(3x^2 + 2x - 6) - (2x + x^2 + 3)$.

Horizontal Method

Use additive inverses to rewrite as addition. Then group like terms.

$(3x^2 + 2x - 6) - (2x + x^2 + 3)$
$= (3x^2 + 2x - 6) + [(-2x) + (-x^2) + (-3)]$
$= [3x^2 + (-x^2)] + [2x + (-2x)] + [-6 + (-3)]$
$= 2x^2 + (-9)$
$= 2x^2 - 9$

The difference is $2x^2 - 9$.

Vertical Method

Align like terms in columns and subtract by adding the additive inverse.

$$\begin{array}{r} 3x^2 + 2x - 6 \\ (-)\quad x^2 + 2x + 3 \\ \hline \end{array}$$

$$\begin{array}{r} 3x^2 + 2x - 6 \\ (+)\; -x^2 - 2x - 3 \\ \hline 2x^2 \qquad\; - 9 \end{array}$$

The difference is $2x^2 - 9$.

Exercises

Find each difference.

1. $(3a - 5) - (5a + 1)$

2. $(9x + 2) - (-3x^2 - 5)$

3. $(9xy + y - 2x) - (6xy - 2x)$

4. $(x^2 + y^2) - (-x^2 + y^2)$

5. $(6p^2 + 4p + 5) - (2p^2 - 5p + 1)$

6. $(6x^2 + 5xy - 2y^2) - (-xy - 2x^2 - 4y^2)$

7. $(8p - 5q) - (-6p^2 + 6q - 3)$

8. $(8x^2 - 4x - 3) - (-2x - x^2 + 5)$

9. $(3x^2 - 2x) - (3x^2 + 5x - 1)$

10. $(4x^2 + 6xy + 2y^2) - (-x^2 + 2xy - 5y^2)$

11. $(2h - 6j - 2k) - (-7h - 5j - 4k)$

12. $(9xy^2 + 5xy) - (-2xy - 8xy^2)$

13. $(2a - 8b) - (-3a + 5b)$

14. $(2x^2 - 8) - (-2x^2 - 6)$

15. $(6z^2 + 4z + 2) - (4z^2 + z)$

16. $(6x^2 - 5x + 1) - (-7x^2 - 2x + 4)$

NAME _____ DATE _____ PERIOD _____

7-5 Study Guide and Intervention

Multiplying a Polynomial by a Monomial

Product of Monomial and Polynomial The Distributive Property can be used to multiply a polynomial by a monomial. You can multiply horizontally or vertically. Sometimes multiplying results in like terms. The products can be simplified by combining like terms.

Example 1 Find $-3x^2(4x^2 + 6x - 8)$.

Horizontal Method
$-3x^2(4x^2 + 6x - 8)$
$= -3x^2(4x^2) + (-3x^2)(6x) - (-3x^2)(8)$
$= -12x^4 + (-18x^3) - (-24x^2)$
$= -12x^4 - 18x^3 + 24x^2$

Vertical Method
$$\begin{array}{r} 4x^2 + 6x - 8 \\ (\times) \quad\quad\quad -3x^2 \\ \hline -12x^4 - 18x^3 + 24x^2 \end{array}$$

The product is $-12x^4 - 18x^3 + 24x^2$.

Example 2 Simplify $-2(4x^2 + 5x) - x(x^2 + 6x)$.

$-2(4x^2 + 5x) - x(x^2 + 6x)$
$= -2(4x^2) + (-2)(5x) + (-x)(x^2) + (-x)(6x)$
$= -8x^2 + (-10x) + (-x^3) + (-6x^2)$
$= (-x^3) + [-8x^2 + (-6x^2)] + (-10x)$
$= -x^3 - 14x^2 - 10x$

Exercises

Find each product.

1. $x(5x + x^2)$

2. $x(4x^2 + 3x + 2)$

3. $-2xy(2y + 4x^2)$

4. $-2g(g^2 - 2g + 2)$

5. $3x(x^4 + x^3 + x^2)$

6. $-4x(2x^3 - 2x + 3)$

7. $-4cx(10 + 3x)$

8. $3y(-4x - 6x^3 - 2y)$

9. $2x^2y^2(3xy + 2y + 5x)$

Simplify.

10. $x(3x - 4) - 5x$

11. $-x(2x^2 - 4x) - 6x^2$

12. $6a(2a - b) + 2a(-4a + 5b)$

13. $4r(2r^2 - 3r + 5) + 6r(4r^2 + 2r + 8)$

14. $4n(3n^2 + n - 4) - n(3 - n)$

15. $2b(b^2 + 4b + 8) - 3b(3b^2 + 9b - 18)$

16. $-2z(4z^2 - 3z + 1) - z(3z^2 + 2z - 1)$

17. $2(4x^2 - 2x) - 3(-6x^2 + 4) + 2x(x - 1)$

7-5 Study Guide and Intervention (continued)

Multiplying a Polynomial by a Monomial

Solve Equations with Polynomial Expressions Many equations contain polynomials that must be added, subtracted, or multiplied before the equation can be solved.

Example Solve $4(n - 2) + 5n = 6(3 - n) + 19$.

$4(n - 2) + 5n = 6(3 - n) + 19$	Original equation
$4n - 8 + 5n = 18 - 6n + 19$	Distributive Property
$9n - 8 = 37 - 6n$	Combine like terms.
$15n - 8 = 37$	Add 6n to both sides.
$15n = 45$	Add 8 to both sides.
$n = 3$	Divide each side by 15.

The solution is 3.

Exercises

Solve each equation.

1. $2(a - 3) = 3(-2a + 6)$

2. $3(x + 5) - 6 = 18$

3. $3x(x - 5) - 3x^2 = -30$

4. $6(x^2 + 2x) = 2(3x^2 + 12)$

5. $4(2p + 1) - 12p = 2(8p + 12)$

6. $2(6x + 4) + 2 = 4(x - 4)$

7. $-2(4y - 3) - 8y + 6 = 4(y - 2)$

8. $c(c + 2) - c(c - 6) = 10c - 12$

9. $3(x^2 - 2x) = 3x^2 + 5x - 11$

10. $2(4x + 3) + 2 = -4(x + 1)$

11. $3(2h - 6) - (2h + 1) = 9$

12. $3(y + 5) - (4y - 8) = -2y + 10$

13. $3(2a - 6) - (-3a - 1) = 4a - 2$

14. $5(2x^2 - 1) - (10x^2 - 6) = -(x + 2)$

15. $3(x + 2) + 2(x + 1) = -5(x - 3)$

16. $4(3p^2 + 2p) - 12p^2 = 2(8p + 6)$

NAME _____ DATE _____ PERIOD ____

7-6 Study Guide and Intervention

Multiplying Polynomials

Multiply Binomials To multiply two binomials, you can apply the Distributive Property twice. A useful way to keep track of terms in the product is to use the FOIL method as illustrated in Example 2.

Example 1 Find $(x + 3)(x - 4)$.

Horizontal Method
$(x + 3)(x - 4)$
$= x(x - 4) + 3(x - 4)$
$= (x)(x) + x(-4) + 3(x) + 3(-4)$
$= x^2 - 4x + 3x - 12$
$= x^2 - x - 12$

Vertical Method

$$\begin{array}{r} x + 3 \\ (\times) \ x - 4 \\ \hline -4x - 12 \\ x^2 + 3x \\ \hline x^2 - x - 12 \end{array}$$

The product is $x^2 - x - 12$.

Example 2 Find $(x - 2)(x + 5)$ using the FOIL method.

$(x - 2)(x + 5)$
       First   Outer   Inner   Last
$= (x)(x) + (x)(5) + (-2)(x) + (-2)(5)$
$= x^2 + 5x + (-2x) - 10$
$= x^2 + 3x - 10$

The product is $x^2 + 3x - 10$.

Exercises

Find each product.

1. $(x + 2)(x + 3)$

2. $(x - 4)(x + 1)$

3. $(x - 6)(x - 2)$

4. $(p - 4)(p + 2)$

5. $(y + 5)(y + 2)$

6. $(2x - 1)(x + 5)$

7. $(3n - 4)(3n - 4)$

8. $(8m - 2)(8m + 2)$

9. $(k + 4)(5k - 1)$

10. $(3x + 1)(4x + 3)$

11. $(x - 8)(-3x + 1)$

12. $(5t + 4)(2t - 6)$

13. $(5m - 3n)(4m - 2n)$

14. $(a - 3b)(2a - 5b)$

15. $(8x - 5)(8x + 5)$

16. $(2n - 4)(2n + 5)$

17. $(4m - 3)(5m - 5)$

18. $(7g - 4)(7g + 4)$

NAME _____ DATE _____ PERIOD _____

7-6 Study Guide and Intervention (continued)
Multiplying Polynomials

Multiply Polynomials The Distributive Property can be used to multiply any two polynomials.

Example Find $(3x + 2)(2x^2 - 4x + 5)$.

$(3x + 2)(2x^2 - 4x + 5)$
$= 3x(2x^2 - 4x + 5) + 2(2x^2 - 4x + 5)$ Distributive Property
$= 6x^3 - 12x^2 + 15x + 4x^2 - 8x + 10$ Distributive Property
$= 6x^3 - 8x^2 + 7x + 10$ Combine like terms.

The product is $6x^3 - 8x^2 + 7x + 10$.

Exercises

Find each product.

1. $(x + 2)(x^2 - 2x + 1)$

2. $(x + 3)(2x^2 + x - 3)$

3. $(2x - 1)(x^2 - x + 2)$

4. $(p - 3)(p^2 - 4p + 2)$

5. $(3k + 2)(k^2 + k - 4)$

6. $(2t + 1)(10t^2 - 2t - 4)$

7. $(3n - 4)(n^2 + 5n - 4)$

8. $(8x - 2)(3x^2 + 2x - 1)$

9. $(2a + 4)(2a^2 - 8a + 3)$

10. $(3x - 4)(2x^2 + 3x + 3)$

11. $(n^2 + 2n - 1)(n^2 + n + 2)$

12. $(t^2 + 4t - 1)(2t^2 - t - 3)$

13. $(y^2 - 5y + 3)(2y^2 + 7y - 4)$

14. $(3b^2 - 2b + 1)(2b^2 - 3b - 4)$

Study Guide and Intervention   Glencoe Algebra 1

NAME _____ DATE _____ PERIOD _____

7-7 Study Guide and Intervention

Special Products

Squares of Sums and Differences Some pairs of binomials have products that follow specific patterns. One such pattern is called the *square of a sum*. Another is called the *square of a difference*.

Square of a sum	$(a + b)^2 = (a + b)(a + b) = a^2 + 2ab + b^2$
Square of a difference	$(a - b)^2 = (a - b)(a - b) = a^2 - 2ab + b^2$

Example 1 Find $(3a + 4)(3a + 4)$.

Use the square of a sum pattern, with $a = 3a$ and $b = 4$.

$(3a + 4)(3a + 4) = (3a)^2 + 2(3a)(4) + (4)^2$
$= 9a^2 + 24a + 16$

The product is $9a^2 + 24a + 16$.

Example 2 Find $(2z - 9)(2z - 9)$.

Use the square of a difference pattern with $a = 2z$ and $b = 9$.

$(2z - 9)(2z - 9) = (2z)^2 - 2(2z)(9) + (9)(9)$
$= 4z^2 - 36z + 81$

The product is $4z^2 - 36z + 81$.

Exercises

Find each product.

1. $(x - 6)^2$

2. $(3p + 4)^2$

3. $(4x - 5)^2$

4. $(2x - 1)^2$

5. $(2h + 3)^2$

6. $(m + 5)^2$

7. $(c + 3)^2$

8. $(3 - p)^2$

9. $(x - 5y)^2$

10. $(8y + 4)^2$

11. $(8 + x)^2$

12. $(3a - 2b)^2$

13. $(2x - 8)^2$

14. $(x^2 + 1)^2$

15. $(m^2 - 2)^2$

16. $(x^3 - 1)^2$

17. $(2h^2 - k^2)^2$

18. $\left(\frac{1}{4}x + 3\right)^2$

19. $(x - 4y^2)^2$

20. $(2p + 4q)^2$

21. $\left(\frac{2}{3}x - 2\right)^2$

Study Guide and Intervention   Glencoe Algebra 1

7-7 Study Guide and Intervention (continued)

Special Products

Product of a Sum and a Difference There is also a pattern for the product of a sum and a difference of the same two terms, $(a + b)(a - b)$. The product is called the **difference of squares**.

Product of a Sum and a Difference	$(a + b)(a - b) = a^2 - b^2$

Example Find $(5x + 3y)(5x - 3y)$.

$(a + b)(a - b) = a^2 - b^2$ Product of a Sum and a Difference
$(5x + 3y)(5x - 3y) = (5x)^2 - (3y)^2$ $a = 5x$ and $b = 3y$
$\qquad\qquad\qquad\quad = 25x^2 - 9y^2$ Simplify.

The product is $25x^2 - 9y^2$.

Exercises

Find each product.

1. $(x - 4)(x + 4)$
2. $(p + 2)(p - 2)$
3. $(4x - 5)(4x + 5)$

4. $(2x - 1)(2x + 1)$
5. $(h + 7)(h - 7)$
6. $(m - 5)(m + 5)$

7. $(2c - 3)(2c + 3)$
8. $(3 - 5q)(3 + 5q)$
9. $(x - y)(x + y)$

10. $(y - 4x)(y + 4x)$
11. $(8 + 4x)(8 - 4x)$
12. $(3a - 2b)(3a + 2b)$

13. $(3y - 8)(3y + 8)$
14. $(x^2 - 1)(x^2 + 1)$
15. $(m^2 - 5)(m^2 + 5)$

16. $(x^3 - 2)(x^3 + 2)$
17. $(h^2 - k^2)(h^2 + k^2)$
18. $\left(\frac{1}{4}x + 2\right)\left(\frac{1}{4}x - 2\right)$

19. $(3x - 2y^2)(3x + 2y^2)$
20. $(2p - 5s)(2p + 5s)$
21. $\left(\frac{4}{3}x - 2y\right)\left(\frac{4}{3}x + 2y\right)$

NAME _____ DATE _____ PERIOD _____

8-1 Study Guide and Intervention

Monomials and Factoring

Prime Factorization When two or more numbers are multiplied, each number is called a **factor** of the product.

	Definition	Example
Prime Number	A prime number is a whole number, greater than 1, whose only factors are 1 and itself.	5
Composite Number	A composite number is a whole number, greater than 1, that has more than two factors.	10
Prime Factorization	Prime factorization occurs when a whole number is expressed as a product of factors that are all prime numbers.	$45 = 3^2 \cdot 5$

Example 1 Factor each number. Then classify each number as *prime* or *composite*.

a. **28**

To find the factors of 28, list all pairs of whole numbers whose product is 28.

$1 \times 28 \quad 2 \times 14 \quad 4 \times 7$

Therefore, the factors of 28 are 1, 2, 4, 7, 14, and 28. Since 28 has more than 2 factors, it is a composite number.

b. **31**

To find the factors of 31, list all pairs of whole numbers whose product is 31.

1×31

Therefore, the factors of 31 are 1 and 31. Since the only factors of 31 are itself and 1, it is a prime number.

Example 2 Find the prime factorization of 200.

Method 1

$200 = 2 \cdot 100$
$ = 2 \cdot 2 \cdot 50$
$ = 2 \cdot 2 \cdot 2 \cdot 25$
$ = 2 \cdot 2 \cdot 2 \cdot 5 \cdot 5$

All the factors in the last row are prime, so the prime factorization of 200 is $2^3 \cdot 5^2$.

Method 2

Use a factor tree.

All of the factors in each last branch of the factor tree are prime, so the prime factorization of 200 is $2^3 \cdot 5^2$.

Exercises

Find the factors of each number. Then classify the number as *prime* or *composite*.

1. 41

2. 121

3. 90

4. 2865

Find the prime factorization of each integer.

5. 600

6. 175

7. −150

Factor each monomial completely.

8. $32x^2$

9. $18m^2n$

10. $49a^3b^2$

Study Guide and Intervention · · · Glencoe Algebra 1

NAME _____ DATE _____ PERIOD _____

8-1 Study Guide and Intervention (continued)

Monomials and Factoring

Greatest Common Factor

Greatest Common Factor (GCF)	
Integers	the greatest number that is a factor of all the integers
Monomials	the product of their common factors when each monomial is expressed in factored form

If two or more integers or monomials have no common prime factors, their GCF is 1 and the integers or monomials are said to be **relatively prime**.

Example Find the GCF of each set of monomials.

a. 12 and 18

$12 = ②\cdot 2 \cdot ③$ Factor each number.
$18 = ②\cdot ③\cdot 3$ Circle the common prime factors, if any.
The GCF of 12 and 18 is $2 \cdot 3$ or 6.

b. $16xy^2z^2$ and $72xyz^3$

$16xy^2z^2 = ②\cdot ②\cdot ②\cdot 2 \cdot ⓧ\cdot ⓨ\cdot y \cdot ⓩ\cdot ⓩ$
$72xyz^3 = ②\cdot ②\cdot ②\cdot 3 \cdot 3 \cdot ⓧ\cdot ⓨ\cdot ⓩ\cdot ⓩ\cdot z$
The GCF of $16xy^2z^2$ and $72xyz^3$ is $2 \cdot 2 \cdot 2 \cdot x \cdot y \cdot z \cdot z$ or $8xyz^2$.

Exercises

Find the GCF of each set of monomials.

1. 12, 48
2. 18, 42
3. 64, 80
4. 32, 54
5. 27, 32
6. 44, 100
7. 45, 15
8. 169, 13
9. 20, 440
10. $49x, 343x^2$
11. $4a^7b, 28ab$
12. $96y, 12x, 8y$
13. $12a, 18abc$
14. $28y^2, 35xy, 49x^2yz$
15. $2m^2n, 12mn^2, 18mn$
16. $12x^2, 32x^2yz, 60xy^2$
17. $18a^3b^2, 36a^3b^2$
18. $15mn^2, 30m^3n^2, 90m^3$
19. $2x^2y, 9x^2y^3, 18xy^2$
20. $a^4b, 8a^3b^2$
21. $ab^2, 5a^4b^2, 10b^3$

Study Guide and Intervention   102   Glencoe Algebra 1

NAME _____ DATE _____ PERIOD _____

8-2 Study Guide and Intervention

Factoring Using the Distributive Property

Factor by Using the Distributive Property The Distributive Property has been used to multiply a polynomial by a monomial. It can also be used to express a polynomial in factored form. Compare the two columns in the table below.

Multiplying	Factoring
$3(a + b) = 3a + 3b$	$3a + 3b = 3(a + b)$
$x(y - z) = xy - xz$	$xy - xz = x(y - z)$
$6y(2x + 1) = 6y(2x) + 6y(1)$ $= 12xy + 6y$	$12xy + 6y = 6y(2x) + 6y(1)$ $= 6y(2x + 1)$

Example 1 Use the Distributive Property to factor $12mn + 80m^2$.

Find the GCF of $12mn$ and $80m^2$.
$12mn = 2 \cdot 2 \cdot 3 \cdot m \cdot n$
$80m^2 = 2 \cdot 2 \cdot 2 \cdot 2 \cdot 5 \cdot m \cdot m$
GCF $= 2 \cdot 2 \cdot m$ or $4m$

Write each term as the product of the GCF and its remaining factors.

$12mn + 80m^2 = 4m(3 \cdot n) + 4m(2 \cdot 2 \cdot 5 \cdot m)$
$ = 4m(3n) + 4m(20m)$
$ = 4m(3n + 20m)$

Thus $12mn + 80m^2 = 4m(3n + 20m)$.

Example 2 Factor $6ax + 3ay + 2bx + by$ by grouping.

$6ax + 3ay + 2bx + by$
$= (6ax + 3ay) + (2bx + by)$
$= 3a(2x + y) + b(2x + y)$
$= (3a + b)(2x + y)$

Check using the FOIL method.
$(3a + b)(2x + y)$
$= 3a(2x) + (3a)(y) + (b)(2x) + (b)(y)$
$= 6ax + 3ay + 2bx + by$ ✓

Exercises

Factor each polynomial.

1. $24x + 48y$

2. $30mn^2 + m^2n - 6n$

3. $q^4 - 18q^3 + 22q$

4. $9x^2 - 3x$

5. $4m + 6n - 8mn$

6. $45s^3 - 15s^2$

7. $14c^3 - 42c^5 - 49c^4$

8. $55p^2 - 11p^4 + 44p^5$

9. $14y^3 - 28y^2 + y$

10. $4x + 12x^2 + 16x^3$

11. $4a^2b + 28ab^2 + 7ab$

12. $6y + 12x - 8z$

13. $x^2 + 2x + x + 2$

14. $6y^2 - 4y + 3y - 2$

15. $4m^2 + 4mn + 3mn + 3n^2$

16. $12ax + 3xz + 4ay + yz$

17. $12a^2 + 3a - 8a - 2$

18. $xa + ya + x + y$

NAME _____ DATE _____ PERIOD ____

8-2 Study Guide and Intervention (continued)
Factoring Using the Distributive Property

Solve Equations by Factoring The following property, along with factoring, can be used to solve certain equations.

Zero Product Property	For any real numbers a and b, if $ab = 0$, then either $a = 0$, $b = 0$, or both a and b equal 0.

Example Solve $9x^2 + x = 0$. Then check the solutions.

Write the equation so that it is of the form $ab = 0$.

$9x^2 + x = 0$ Original equation
$x(9x + 1) = 0$ Factor the GCF of $9x^2 + x$, which is x.
$x = 0$ or $9x + 1 = 0$ Zero Product Property
$x = 0$ $x = -\dfrac{1}{9}$ Solve each equation.

The solution set is $\left\{0, -\dfrac{1}{9}\right\}$.

CHECK Substitute 0 and $-\dfrac{1}{9}$ for x in the original equation.

$9x^2 + x = 0$ $9x^2 + x = 0$
$9(0)^2 + 0 = 0$ $9\left(-\dfrac{1}{9}\right)^2 + \left(-\dfrac{1}{9}\right) = 0$
$0 = 0 \checkmark$ $\dfrac{1}{9} + \left(-\dfrac{1}{9}\right) = 0$
 $0 = 0 \checkmark$

Exercises

Solve each equation. Check your solutions.

1. $x(x + 3) = 0$
2. $3m(m - 4) = 0$
3. $(r - 3)(r + 2) = 0$

4. $3x(2x - 1) = 0$
5. $(4m + 8)(m - 3) = 0$
6. $5s^2 = 25s$

7. $(4c + 2)(2c - 7) = 0$
8. $5p - 15p^2 = 0$
9. $4y^2 = 28y$

10. $12x^2 = -6x$
11. $(4a + 3)(8a + 7) = 0$
12. $8y = 12y^2$

13. $x^2 = -2x$
14. $(6y - 4)(y + 3) = 0$
15. $4m^2 = 4m$

16. $12x = 3x^2$
17. $12a^2 = -3a$
18. $(12a + 4)(3a - 1) = 0$

8-3 Study Guide and Intervention

Factoring Trinomials: $x^2 + bx + c$

Factor $x^2 + bx + c$ To factor a trinomial of the form $x^2 + bx + c$, find two integers, m and n, whose sum is equal to b and whose product is equal to c.

Factoring $x^2 + bx + c$	$x^2 + bx + c = (x + m)(x + n)$, where $m + n = b$ and $mn = c$.

Example 1 Factor each trinomial.

a. $x^2 + 7x + 10$

In this trinomial, $b = 7$ and $c = 10$.

Factors of 10	Sum of Factors
1, 10	11
2, 5	7

Since $2 + 5 = 7$ and $2 \cdot 5 = 10$, let $m = 2$ and $n = 5$.
$x^2 + 7x + 10 = (x + 5)(x + 2)$

b. $x^2 - 8x + 7$

In this trinomial, $b = -8$ and $c = 7$. Notice that $m + n$ is negative and mn is positive, so m and n are both negative. Since $-7 + (-1) = -8$ and $(-7)(-1) = 7$, $m = -7$ and $n = -1$.
$x^2 - 8x + 7 = (x - 7)(x - 1)$

Example 2 Factor $x^2 + 6x - 16$.

In this trinomial, $b = 6$ and $c = -16$. This means $m + n$ is positive and mn is negative. Make a list of the factors of -16, where one factor of each pair is positive.

Factors of -16	Sum of Factors
1, -16	-15
-1, 16	15
2, -8	-6
-2, 8	6

Therefore, $m = -2$ and $n = 8$.
$x^2 + 6x - 16 = (x - 2)(x + 8)$

Exercises

Factor each trinomial.

1. $x^2 + 4x + 3$
2. $m^2 + 12m + 32$
3. $r^2 - 3r + 2$

4. $x^2 - x - 6$
5. $x^2 - 4x - 21$
6. $x^2 - 22x + 121$

7. $c^2 - 4c - 12$
8. $p^2 - 16p + 64$
9. $9 - 10x + x^2$

10. $x^2 + 6x + 5$
11. $a^2 + 8a - 9$
12. $y^2 - 7y - 8$

13. $x^2 - 2x - 3$
14. $y^2 + 14y + 13$
15. $m^2 + 9m + 20$

16. $x^2 + 12x + 20$
17. $a^2 - 14a + 24$
18. $18 + 11y + y^2$

19. $x^2 + 2xy + y^2$
20. $a^2 - 4ab + 4b^2$
21. $x^2 + 6xy - 7y^2$

8-3 Study Guide and Intervention (continued)

Factoring Trinomials: $x^2 + bx + c$

Solve Equations by Factoring Factoring and the Zero Product Property from Lesson 9-2 can be used to solve many equations of the form $x^2 + bx + c = 0$.

Example 1 Solve $x^2 + 6x = 7$. Check your solutions.

$x^2 + 6x = 7$	Original equation
$x^2 + 6x - 7 = 0$	Rewrite equation so that one side equals 0.
$(x - 1)(x + 7) = 0$	Factor.
$x - 1 = 0$ or $x + 7 = 0$	Zero Product Property
$x = 1 \quad\quad x = -7$	Solve each equation.

The solution set is {1, −7}. Since $1^2 + 6 = 7$ and $(-7)^2 + 6(-7) = 7$, the solutions check.

Example 2 ROCKET LAUNCH A rocket is fired with an initial velocity of 2288 feet per second. How many seconds will it take for the rocket to hit the ground?

The formula $h = vt - 16t^2$ gives the height h of the rocket after t seconds when the initial velocity v is given in feet per second.

$h = vt - 16t^2$	Formula
$0 = 2288t - 16t^2$	Substitute.
$0 = 16t(143 - t)$	Factor.
$16t = 0$ or $143 - t = 0$	Zero Product Property
$t = 0 \quad\quad t = 143$	Solve each equation.

The value $t = 0$ represents the time at launch. The rocket returns to the ground in 143 seconds, or a little less than 2.5 minutes after launch.

Exercises

Solve each equation. Check your solutions.

1. $x^2 - 4x + 3 = 0$
2. $y^2 - 5y + 4 = 0$
3. $m^2 + 10m + 9 = 0$

4. $x^2 = x + 2$
5. $x^2 - 4x = 5$
6. $x^2 - 12x + 36 = 0$

7. $c^2 - 8 = -7c$
8. $p^2 = 9p - 14$
9. $-9 - 8x + x^2 = 0$

10. $x^2 + 6 = 5x$
11. $a^2 = 11a - 18$
12. $y^2 - 8y + 15 = 0$

13. $x^2 = 24 - 10x$
14. $a^2 - 18a = -72$
15. $b^2 = 10b - 16$

Use the formula $h = vt - 16t^2$ to solve each problem.

16. **FOOTBALL** A punter can kick a football with an initial velocity of 48 feet per second. How many seconds will it take for the ball to return to the ground?

17. **BASEBALL** A ball is thrown up with an initial velocity of 32 feet per second. How many seconds will it take for the ball to return to the ground?

18. **ROCKET LAUNCH** If a rocket is launched with an initial velocity of 1600 feet per second, when will the rocket be 14,400 feet high?

NAME _____ DATE _____ PERIOD _____

8-4 Study Guide and Intervention

Factoring Trinomials: $ax^2 + bx + c$

Factor $ax^2 + bx + c$ To factor a trinomial of the form $ax^2 + bx + c$, find two integers, m and n whose product is equal to ac and whose sum is equal to b. If there are no integers that satisfy these requirements, the polynomial is called a **prime polynomial**.

Example 1 Factor $2x^2 + 15x + 18$.

In this example, $a = 2$, $b = 15$, and $c = 18$. You need to find two numbers whose sum is 15 and whose product is $2 \cdot 18$ or 36. Make a list of the factors of 36 and look for the pair of factors whose sum is 15.

Factors of 36	Sum of Factors
1, 36	37
2, 18	20
3, 12	15

Use the pattern $ax^2 + mx + nx + c$, with $a = 2$, $m = 3$, $n = 12$, and $c = 18$.

$2x^2 + 15x + 18 = 2x^2 + 3x + 12x + 18$
$= (2x^2 + 3x) + (12x + 18)$
$= x(2x + 3) + 6(2x + 3)$
$= (x + 6)(2x + 3)$

Therefore, $2x^2 + 15x + 18 = (x + 6)(2x + 3)$.

Example 2 Factor $3x^2 - 3x - 18$.

Note that the GCF of the terms $3x^2$, $3x$, and 18 is 3. First factor out this GCF.

$3x^2 - 3x - 18 = 3(x^2 - x - 6)$.

Now factor $x^2 - x - 6$. Since $a = 1$, find the two factors of -6 whose sum is -1.

Factors of -6	Sum of Factors
1, -6	-5
-1, 6	5
-2, 3	1
2, -3	-1

Now use the pattern $(x + m)(x + n)$ with $m = 2$ and $n = -3$.

$x^2 - x - 6 = (x + 2)(x - 3)$

The complete factorization is
$3x^2 - 3x - 18 = 3(x + 2)(x - 3)$.

Exercises

Factor each trinomial, if possible. If the trinomial cannot be factored using integers, write *prime*.

1. $2x^2 - 3x - 2$
2. $3m^2 - 8m - 3$
3. $16r^2 - 8r + 1$

4. $6x^2 + 5x - 6$
5. $3x^2 + 2x - 8$
6. $18x^2 - 27x - 5$

7. $2a^2 + 5a + 3$
8. $18y^2 + 9y - 5$
9. $-4c^2 + 19c - 21$

10. $8x^2 - 4x - 24$
11. $28p^2 + 60p - 25$
12. $48x^2 + 22x - 15$

13. $3y^2 - 6y - 24$
14. $4x^2 + 26x - 48$
15. $8m^2 - 44m + 48$

16. $6x^2 - 7x + 18$
17. $2a^2 - 14a + 18$
18. $18 + 11y + 2y^2$

Study Guide and Intervention   Glencoe Algebra 1

NAME _____ DATE _____ PERIOD _____

8-4 Study Guide and Intervention (continued)

Factoring Trinomials: $ax^2 + bx + c$

Solve Equations by Factoring Factoring and the Zero Product Property can be used to solve some equations of the form $ax^2 + bx + c = 0$.

Example Solve $12x^2 + 3x = 2 - 2x$. Check your solutions.

$12x^2 + 3x = 2 - 2x$	Original equation
$12x^2 + 5x - 2 = 0$	Rewrite equation so that one side equals 0.
$(3x + 2)(4x - 1) = 0$	Factor the left side.
$3x + 2 = 0$ or $4x - 1 = 0$	Zero Product Property
$x = -\frac{2}{3}$ $x = \frac{1}{4}$	Solve each equation.

The solution set is $\left\{-\frac{2}{3}, \frac{1}{4}\right\}$.

Since $12\left(-\frac{2}{3}\right)^2 + 3\left(-\frac{2}{3}\right) = 2 - 2\left(-\frac{2}{3}\right)$ and $12\left(\frac{1}{4}\right)^2 + 3\left(\frac{1}{4}\right) = 2 - 2\left(\frac{1}{4}\right)$, the solutions check.

Exercises

Solve each equation. Check your solutions.

1. $8x^2 + 2x - 3 = 0$

2. $3n^2 - 2n - 5 = 0$

3. $2d^2 - 13d - 7 = 0$

4. $4x^2 = x + 3$

5. $3x^2 - 13x = 10$

6. $6x^2 - 11x - 10 = 0$

7. $2k^2 - 40 = -11k$

8. $2p^2 = -21p - 40$

9. $-7 - 18x + 9x^2 = 0$

10. $12x^2 - 15 = -8x$

11. $7a^2 = -65a - 18$

12. $16y^2 - 2y - 3 = 0$

13. $8x^2 + 5x = 3 + 7x$

14. $4a^2 - 18a + 5 = 15$

15. $3b^2 - 18b = 10b - 49$

16. The difference of the squares of two consecutive odd integers is 24. Find the integers.

17. **GEOMETRY** The length of a Charlotte, North Carolina, conservatory garden is 20 yards greater than its width. The area is 300 square yards. What are the dimensions?

18. **GEOMETRY** A rectangle with an area of 24 square inches is formed by cutting strips of equal width from a rectangular piece of paper. Find the dimensions of the new rectangle if the original rectangle measures 8 inches by 6 inches.

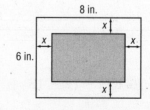

8-5 Study Guide and Intervention

Factoring Differences of Squares

Factor $a^2 - b^2$ The binomial expression $a^2 - b^2$ is called the **difference of two squares**. The following pattern shows how to factor the difference of squares.

Difference of Squares	$a^2 - b^2 = (a - b)(a + b) = (a + b)(a - b)$.

Example 1 Factor each binomial.

a. $n^2 - 64$
$n^2 - 64$
$= n^2 - 8^2$ Write in the form $a^2 - b^2$.
$= (n + 8)(n - 8)$ Factor.

b. $4m^2 - 81n^2$
$4m^2 - 81n^2$
$= (2m)^2 - (9n)^2$ Write in the form $a^2 - b^2$.
$= (2m - 9n)(2m + 9n)$ Factor.

Example 2 Factor each polynomial.

a. $50a^2 - 72$
$50a^2 - 72$
$= 2(25a^2 - 36)$ Find the GCF.
$= 2[(5a)^2 - 6^2)]$ $25a^2 = 5a \cdot 5a$ and $36 = 6 \cdot 6$
$= 2(5a + 6)(5a - 6)$ Factor the difference of squares.

b. $4x^4 + 8x^3 - 4x^2 - 8x$
$4x^4 + 8x^3 - 4x^2 - 8x$ Original polynomial
$= 4x(x^3 + 2x^2 - x - 2)$ Find the GCF.
$= 4x[(x^3 + 2x^2) - (x + 2)]$ Group terms.
$= 4x[x^2(x + 2) - 1(x + 2)]$ Find the GCF.
$= 4x[(x^2 - 1)(x + 2)]$ Factor by grouping.
$= 4x[(x - 1)(x + 1)(x + 2)]$ Factor the difference of squares.

Exercises

Factor each polynomial if possible. If the polynomial cannot be factored, write *prime*.

1. $x^2 - 81$

2. $m^2 - 100$

3. $16n^2 - 25$

4. $36x^2 - 100y^2$

5. $49x^2 - 32$

6. $16a^2 - 9b^2$

7. $225c^2 - a^2$

8. $72p^2 - 50$

9. $-2 + 2x^2$

10. $-81 + a^4$

11. $6 - 54a^2$

12. $8y^2 - 200$

13. $4x^3 - 100x$

14. $2y^4 - 32y^2$

15. $8m^3 - 128m$

16. $6x^2 - 25$

17. $2a^3 - 98ab^2$

18. $18y^2 - 72y^4$

19. $169x^3 - x$

20. $3a^4 - 3a^2$

21. $3x^4 + 6x^3 - 3x^2 - 6x$

8-5 Study Guide and Intervention (continued)
Factoring Differences of Squares

Solve Equations by Factoring Factoring and the Zero Product Property can be used to solve equations that can be written as the product of any number of factors set equal to 0.

Example Solve each equation. Check your solutions.

a. $x^2 - \dfrac{1}{25} = 0$

$\qquad x^2 - \dfrac{1}{25} = 0 \qquad$ Original equation

$\qquad x^2 - \left(\dfrac{1}{5}\right)^2 = 0 \qquad x^2 = x \cdot x$ and $\dfrac{1}{25} = \left(\dfrac{1}{5}\right)\left(\dfrac{1}{5}\right)$

$\qquad \left(x + \dfrac{1}{5}\right)\left(x - \dfrac{1}{5}\right) = 0 \qquad$ Factor the difference of squares.

$\qquad x + \dfrac{1}{5} = 0 \quad$ or $\quad x - \dfrac{1}{5} = 0 \qquad$ Zero Product Property

$\qquad x = -\dfrac{1}{5} \qquad\qquad x = \dfrac{1}{5} \qquad$ Solve each equation.

The solution set is $\left\{-\dfrac{1}{5}, \dfrac{1}{5}\right\}$. Since $\left(-\dfrac{1}{5}\right)^2 - \dfrac{1}{25} = 0$ and $\left(\dfrac{1}{5}\right)^2 - \dfrac{1}{25} = 0$, the solutions check.

b. $4x^3 = 9x$

$\qquad 4x^3 = 9x \qquad$ Original equation
$\qquad 4x^3 - 9x = 0 \qquad$ Subtract 9x from each side.
$\qquad x(4x^2 - 9) = 0 \qquad$ Find the GCF.
$\qquad x[(2x)^2 - 3^2] = 0 \qquad 4x^2 = 2x \cdot 2x$ and $9 = 3 \cdot 3$
$\qquad x[(2x)^2 - 3^2] = x[(2x - 3)(2x + 3)] \qquad$ Factor the difference of squares.
$\qquad x = 0 \quad$ or $\quad (2x - 3) = 0 \quad$ or $\quad (2x + 3) = 0 \qquad$ Zero Product Property
$\qquad x = 0 \qquad\qquad x = \dfrac{3}{2} \qquad\qquad x = -\dfrac{3}{2} \qquad$ Solve each equation.

The solution set is $\left\{0, \dfrac{3}{2}, -\dfrac{3}{2}\right\}$.

Since $4(0)^3 = 9(0)$, $4\left(\dfrac{3}{2}\right)^3 = 9\left(\dfrac{3}{2}\right)$, and $4\left(-\dfrac{3}{2}\right)^3 = 9\left(-\dfrac{3}{2}\right)$, the solutions check.

Exercises

Solve each equation. Check your solutions.

1. $81x^2 = 49$

2. $36n^2 = 1$

3. $25d^2 - 100 = 0$

4. $\dfrac{1}{4}x^2 = 25$

5. $36 = \dfrac{1}{25}x^2$

6. $\dfrac{49}{100} - x^2 = 0$

7. $9x^3 = 25x$

8. $7a^3 = 175a$

9. $2m^3 = 32m$

10. $16y^3 = 25y$

11. $\dfrac{1}{64}x^2 = 49$

12. $4a^3 - 64a = 0$

13. $3b^3 - 27b = 0$

14. $\dfrac{9}{25}m^2 = 121$

15. $48n^3 = 147n$

8-6 Study Guide and Intervention

Perfect Squares and Factoring

Factor Perfect Square Trinomials

| Perfect Square Trinomial | a trinomial of the form $a^2 + 2ab + b^2$ or $a^2 - 2ab + b^2$ |

The patterns shown below can be used to factor perfect square trinomials.

Squaring a Binomial	Factoring a Perfect Square Trinomial
$(a + 4)^2 = a^2 + 2(a)(4) + 4^2$ $= a^2 + 8a + 16$	$a^2 + 8a + 16 = a^2 + 2(a)(4) + 4^2$ $= (a + 4)^2$
$(2x - 3)^2 = (2x)^2 - 2(2x)(3) + 3^2$ $= 4x^2 - 12x + 9$	$4x^2 - 12x + 9 = (2x)^2 - 2(2x)(3) + 3^2$ $= (2x - 3)^2$

Example 1 Determine whether $16n^2 - 24n + 9$ is a perfect square trinomial. If so, factor it.

Since $16n^2 = (4n)(4n)$, the first term is a perfect square.
Since $9 = 3 \cdot 3$, the last term is a perfect square.
The middle term is equal to $2(4n)(3)$.
Therefore, $16n^2 - 24n + 9$ is a perfect square trinomial.
$16n^2 - 24n + 9 = (4n)^2 - 2(4n)(3) + 3^2$
$= (4n - 3)^2$

Example 2 Factor $16x^2 - 32x + 15$.

Since 15 is not a perfect square, use a different factoring pattern.

$16x^2 - 32x + 15$ Original trinomial
$= 16x^2 + mx + nx + 15$ Write the pattern.
$= 16x^2 - 12x - 20x + 15$ $m = -12$ and $n = -20$
$= (16x^2 - 12x) - (20x - 15)$ Group terms.
$= 4x(4x - 3) - 5(4x - 3)$ Find the GCF.
$= (4x - 5)(4x - 3)$ Factor by grouping.

Therefore $16x^2 - 32x + 15 = (4x - 5)(4x - 3)$.

Exercises

Determine whether each trinomial is a perfect square trinomial. If so, factor it.

1. $x^2 - 16x + 64$
2. $m^2 + 10m + 25$
3. $p^2 + 8p + 64$

Factor each polynomial if possible. If the polynomial cannot be factored, write *prime*.

4. $98x^2 - 200y^2$
5. $x^2 + 22x + 121$
6. $81 + 18s + s^2$

7. $25c^2 - 10c - 1$
8. $169 - 26r + r^2$
9. $7x^2 - 9x + 2$

10. $16m^2 + 48m + 36$
11. $16 - 25a^2$
12. $b^2 - 16b + 256$

13. $36x^2 - 12x + 1$
14. $16a^2 - 40ab + 25b^2$
15. $8m^3 - 64m$

Study Guide and Intervention

Glencoe Algebra 1

8-6 Study Guide and Intervention (continued)

Perfect Squares and Factoring

Solve Equations with Perfect Squares Factoring and the Zero Product Property can be used to solve equations that involve repeated factors. The repeated factor gives just one solution to the equation. You may also be able to use the **square root property** below to solve certain equations.

Square Root Property	For any number $n > 0$, if $x^2 = n$, then $x = \pm\sqrt{n}$.

Example
Solve each equation. Check your solutions.

a. $x^2 - 6x + 9 = 0$

$x^2 - 6x + 9 = 0$ Original equation
$x^2 - 2(3x) + 3^2 = 0$ Recognize a perfect square trinomial.
$(x - 3)(x - 3) = 0$ Factor the perfect square trinomial.
$x - 3 = 0$ Set repeated factor equal to 0.
$x = 3$ Solve.

The solution set is {3}. Since $3^2 - 6(3) + 9 = 0$, the solution checks.

b. $(a - 5)^2 = 64$

$(a - 5)^2 = 64$ Original equation
$a - 5 = \pm\sqrt{64}$ Square Root Property
$a - 5 = \pm 8$ $64 = 8 \cdot 8$
$a = 5 \pm 8$ Add 5 to each side.
$a = 5 + 8$ or $a = 5 - 8$ Separate into 2 equations.
$a = 13$ $a = -3$ Solve each equation.

The solution set is {−3, 13}. Since $(-3 - 5)^2 = 64$ and $(13 - 5)^2 = 64$, the solutions check.

Exercises

Solve each equation. Check your solutions.

1. $x^2 + 4x + 4 = 0$
2. $16n^2 + 16n + 4 = 0$
3. $25d^2 - 10d + 1 = 0$

4. $x^2 + 10x + 25 = 0$
5. $9x^2 - 6x + 1 = 0$
6. $x^2 + x + \frac{1}{4} = 0$

7. $25k^2 + 20k + 4 = 0$
8. $p^2 + 2p + 1 = 49$
9. $x^2 + 4x + 4 = 64$

10. $x^2 - 6x + 9 = 25$
11. $a^2 + 8a + 16 = 1$
12. $16y^2 + 8y + 1 = 0$

13. $(x + 3)^2 = 49$
14. $(y + 6)^2 = 1$
15. $(m - 7)^2 = 49$

16. $(2x + 1)^2 = 1$
17. $(4x + 3)^2 = 25$
18. $(3h - 2)^2 = 4$

19. $(x + 1)^2 = 7$
20. $(y - 3)^2 = 6$
21. $(m - 2)^2 = 5$

NAME _____ DATE _____ PERIOD _____

9-1 Study Guide and Intervention

Graphing Quadratic Functions

Graph Quadratic Functions

| Quadratic Function | a function described by an equation of the form $f(x) = ax^2 + bx + c$, where $a \neq 0$ | Example: $y = 2x^2 + 3x + 8$ |

The parent graph of the family of quadratic fuctions is $y = x^2$. Graphs of quadratic functions have a general shape called a **parabola**. A parabola opens upward and has a **minimum point** when the value of a is positive, and a parabola opens downward and has a **maximum point** when the value of a is negative.

Example 1

a. Use a table of values to graph $y = x^2 - 4x + 1$.

x	y
-1	6
0	1
1	-2
2	-3
3	-2
4	1

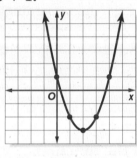

Graph the ordered pairs in the table and connect them with a smooth curve.

b. What is the domain and range of this function?

Example 2

a. Use a table of values to graph $y = -x^2 - 6x - 7$.

x	y
-6	-7
-5	-2
-4	1
-3	2
-2	1
-1	-2
0	-7

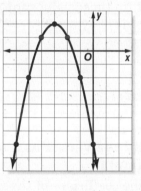

Graph the ordered pairs in the table and connect them with a smooth curve.

b. What is the domain and range of this function?

Exercises

Use a table of values to graph each function. Determine the domain and range.

1. $y = x^2 + 2$

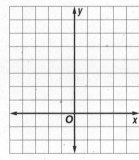

2. $y = -x^2 - 4$

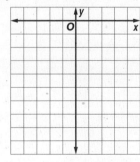

3. $y = x^2 - 3x + 2$

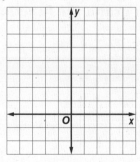

NAME _____ DATE _____ PERIOD _____

9-1 Study Guide and Intervention (continued)

Graphing Quadratic Functions

Symmetry and Vertices Parabolas have a geometric property called **symmetry**. That is, if the figure is folded in half, each half will match the other half exactly. The vertical line containing the fold line is called the **axis of symmetry**.

| Axis of Symmetry | For the parabola $y = ax^2 + bx + c$, where $a \neq 0$, the line $x = -\dfrac{b}{2a}$ is the axis of symmetry. | **Example:** The axis of symmetry of $y = x^2 + 2x + 5$ is the line $x = -1$. |

The axis of symmetry contains the minimum or maximum point of the parabola, the **vertex**.

Example Consider the graph of $y = 2x^2 + 4x + 1$.

a. Write the equation of the axis of symmetry.

In $y = 2x^2 + 4x + 1$, $a = 2$ and $b = 4$. Substitute these values into the equation of the axis of symmetry.

$x = -\dfrac{b}{2a}$

$x = -\dfrac{4}{2(2)} = -1$

The axis of symmetry is $x = -1$.

b. Find the coordinates of the vertex.

Since the equation of the axis of symmetry is $x = -1$ and the vertex lies on the axis, the x-coordinate of the vertex is -1.

$y = 2x^2 + 4x + 1$ Original equation
$y = 2(-1)^2 + 4(-1) + 1$ Substitute.
$y = 2(1) - 4 + 1$ Simplify.
$y = -1$

The vertex is at $(-1, -1)$.

c. Identify the vertex as a maximum or a minimum.

Since the coefficient of the x^2-term is positive, the parabola opens upward, and the vertex is a minimum point.

d. Graph the function.

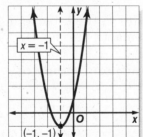

Exercises

Write the equation of the axis of symmetry, and find the coordinates of the vertex of the graph of each function. Identify the vertex as a maximum or a minimum. Then graph the function.

1. $y = x^2 + 3$

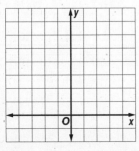

2. $y = -x^2 - 4x - 4$

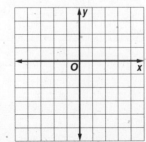

3. $y = x^2 + 2x + 3$

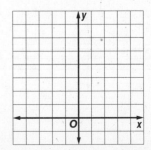

NAME _____ DATE _____ PERIOD _____

9-2 Study Guide and Intervention

Solving Quadratic Equations by Graphing

Solve by Graphing

Quadratic Equation	an equation of the form $ax^2 + bx + c = 0$, where $a \neq 0$

The solutions of a quadratic equation are called the **roots** of the equation. The roots of a quadratic equation can be found by graphing the related quadratic function $f(x) = ax^2 + bx + c$ and finding the x-intercepts or **zeros** of the function.

Example 1 Solve $x^2 + 4x + 3 = 0$ by graphing.

Graph the related function $f(x) = x^2 + 4x + 3$. The equation of the axis of symmetry is $x = -\frac{4}{2(1)}$ or -2. The vertex is at $(-2, -1)$. Graph the vertex and several other points on either side of the axis of symmetry.

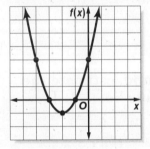

To solve $x^2 + 4x + 3 = 0$, you need to know where the value of $f(x) = 0$. This occurs at the x-intercepts, -3 and -1.
The solutions are -3 and -1.

Example 2 Solve $x^2 - 6x + 9 = 0$ by graphing.

Graph the related function $f(x) = x^2 - 6x + 9$. The equation of the axis of symmetry is $x = \frac{6}{2(1)}$ or 3. The vertex is at $(3, 0)$. Graph the vertex and several other points on either side of the axis of symmetry.

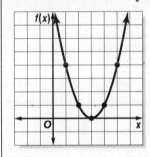

To solve $x^2 - 6x + 9 = 0$, you need to know where the value of $f(x) = 0$. The vertex of the parabola is the x-intercept. Thus, the only solution is 3.

Exercises

Solve each equation by graphing.

1. $x^2 + 7x + 12 = 0$

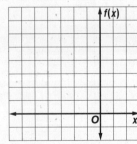

2. $x^2 - x - 12 = 0$

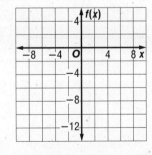

3. $x^2 - 4x + 5 = 0$

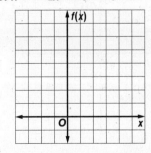

Study Guide and Intervention 115 Glencoe Algebra 1

NAME _____ DATE _____ PERIOD ____

9-2 Study Guide and Intervention (continued)
Solving Quadratic Equations by Graphing

Estimate Solutions The roots of a quadratic equation may not be integers. If exact roots cannot be found, they can be estimated by finding the consecutive integers between which the roots lie.

Example Solve $x^2 + 6x + 6 = 0$ by graphing. If integral roots cannot be found, estimate the roots by stating the consecutive integers between which the roots lie.

Graph the related function $f(x) = x^2 + 6x + 6$.

x	f(x)
-5	1
-4	-2
-3	-3
-2	-2
-1	1

Notice that the value of the function changes from negative to positive between the x-values of -5 and -4 and between -2 and -1.

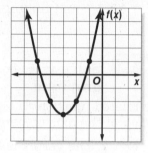

The x-intercepts of the graph are between -5 and -4 and between -2 and -1. So one root is between -5 and -4, and the other root is between -2 and -1.

Exercises

Solve each equation by graphing. If integral roots cannot be found, estimate the roots by stating the consecutive integers between which the roots lie.

1. $x^2 + 7x + 9 = 0$

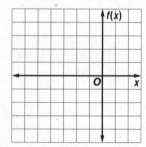

2. $x^2 - x - 4 = 0$

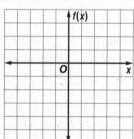

3. $x^2 - 4x + 6 = 0$

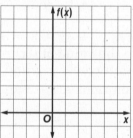

4. $x^2 - 4x - 1 = 0$

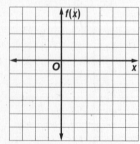

5. $4x^2 - 12x + 3 = 0$

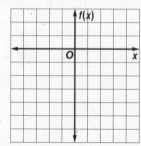

6. $x^2 - 2x - 4 = 0$

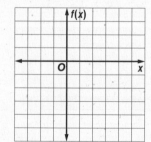

NAME _____ DATE _____ PERIOD ____

9-3 Study Guide and Intervention

Solving Quadratic Equations by Completing the Square

Find the Square Root An equation such as $x^2 - 4x + 4 = 5$ can be solved by taking the square root of each side.

Example 1 Solve $x^2 - 2x + 1 = 9$. Round to the nearest tenth if necessary.

$x^2 - 2x + 1 = 9$
$(x - 1)^2 = 9$
$\sqrt{(x - 1)^2} = \sqrt{9}$
$|x - 1| = \sqrt{9}$
$x - 1 = \pm 3$
$x - 1 + 1 = \pm 3 + 1$
$x = 1 \pm 3$

$x = 1 + 3$ or $x = 1 - 3$
$\quad = 4 \qquad\qquad = -2$

The solution set is $\{-2, 4\}$.

Example 2 Solve $x^2 - 4x + 4 = 5$. Round to the nearest tenth if necessary.

$x^2 - 4x + 4 = 5$
$(x - 2)^2 = 5$
$\sqrt{(x - 2)^2} = \sqrt{5}$
$|x - 2| = \sqrt{5}$
$x - 2 = \pm\sqrt{5}$
$x - 2 + 2 = \pm\sqrt{5} + 2$
$x = 2 \pm \sqrt{5}$

Use a calculator to evaluate each value of x.
$x = 2 + \sqrt{5}$ or $x = 2 - \sqrt{5}$
$\quad \approx 4.2 \qquad\qquad \approx -0.2$

The solution set is $\{-0.2, 4.2\}$.

Exercises

Solve each equation by taking the square root of each side. Round to the nearest tenth if necessary.

1. $x^2 + 4x + 4 = 9$
2. $m^2 + 12m + 36 = 1$
3. $r^2 - 6r + 9 = 16$

4. $x^2 - 2x + 1 = 25$
5. $x^2 - 8x + 16 = 5$
6. $x^2 - 10x + 25 = 8$

7. $c^2 - 4c + 4 = 7$
8. $p^2 + 16p + 64 = 3$
9. $x^2 + 8x + 16 = 9$

10. $x^2 + 6x + 9 = 4$
11. $a^2 + 8a + 16 = 10$
12. $y^2 - 12y + 36 = 5$

13. $x^2 + 10x + 25 = 1$
14. $y^2 + 14y + 49 = 6$
15. $m^2 - 8m + 16 = 2$

16. $x^2 + 12x + 36 = 10$
17. $a^2 - 14a + 49 = 3$
18. $y^2 + 8y + 16 = 7$

Study Guide and Intervention   Glencoe Algebra 1

NAME _____ DATE _____ PERIOD _____

9-3 Study Guide and Intervention (continued)
Solving Quadratic Equations by Completing the Square

Complete the Square Since few quadratic expressions are perfect square trinomials, the method of **completing the square** can be used to solve some quadratic equations. Use the following steps to complete the square for a quadratic expression of the form $ax^2 + bx$.

Step 1	Find $\frac{b}{2}$.
Step 2	Find $\left(\frac{b}{2}\right)^2$.
Step 3	Add $\left(\frac{b}{2}\right)^2$ to $ax^2 + bx$.

Example Solve $x^2 + 6x + 3 = 10$ by completing the square.

$x^2 + 6x + 3 = 10$ Original equation
$x^2 + 6x + 3 - 3 = 10 - 3$ Subtract 3 from each side.
$x^2 + 6x = 7$ Simplify.
$x^2 + 6x + 9 = 7 + 9$ Since $\left(\frac{6}{2}\right)^2 = 9$, add 9 to each side.
$(x + 3)^2 = 16$ Factor $x^2 + 6x + 9$.
$x + 3 = \pm 4$ Take the square root of each side.
$x = -3 \pm 4$ Simplify.

$x = -3 + 4$ or $x = -3 - 4$
$= 1$ $= -7$

The solution set is $\{-7, 1\}$.

Exercises

Solve each equation by completing the square. Round to the nearest tenth if necessary.

1. $t^2 - 4t + 3 = 0$

2. $y^2 + 10y = -9$

3. $y^2 - 8y - 9 = 0$

4. $x^2 - 6x = 16$

5. $p^2 - 4p - 5 = 0$

6. $x^2 - 12x = 9$

7. $c^2 + 8c = 20$

8. $p^2 = 2p + 1$

9. $x^2 + 20x + 11 = -8$

10. $x^2 - 1 = 5x$

11. $a^2 = 22a + 23$

12. $m^2 - 8m = -7$

13. $x^2 + 10x = 24$

14. $a^2 - 18a = 19$

15. $b^2 + 16b = -16$

16. $4x^2 = 24 + 4x$

17. $2m^2 + 4m + 2 = 8$

18. $4k^2 = 40k + 44$

NAME _____ DATE _____ PERIOD _____

9-4 Study Guide and Intervention

Solving Quadratic Equations by Using the Quadratic Formula

Quadratic Formula To solve the standard form of the quadratic equation, $ax^2 + bx + c = 0$, use the **Quadratic Formula**.

Quadratic Formula	the formula $x = \dfrac{-b \pm \sqrt{b^2 - 4ac}}{2a}$ that gives the solutions of $ax^2 + bx + c = 0$, where $a \neq 0$

Example 1 Solve $x^2 + 2x = 3$ by using the Quadratic Formula.

Rewrite the equation in standard form.

$x^2 + 2x = 3$ Original equation
$x^2 + 2x - 3 = 3 - 3$ Subtract 3 from each side.
$x^2 + 2x - 3 = 0$ Simplify.

Now let $a = 1$, $b = 2$, and $c = -3$ in the Quadratic Formula.

$x = \dfrac{-b \pm \sqrt{b^2 - 4ac}}{2a}$

$= \dfrac{-2 \pm \sqrt{(2)^2 - 4(1)(-3)}}{2(1)}$

$= \dfrac{-2 \pm \sqrt{16}}{2}$

$x = \dfrac{-2 + 4}{2}$ or $x = \dfrac{-2 - 4}{2}$

$= 1$ $= -3$

The solution set is $\{-3, 1\}$.

Example 2 Solve $x^2 - 6x - 2 = 0$ by using the Quadratic Formula. Round to the nearest tenth if necessary.

For this equation $a = 1$, $b = -6$, and $c = -2$.

$x = \dfrac{-b \pm \sqrt{b^2 - 4ac}}{2a}$

$= \dfrac{6 \pm \sqrt{(-6)^2 - 4(1)(-2)}}{2(1)}$

$= \dfrac{6 \pm \sqrt{44}}{2}$

$x = \dfrac{6 + \sqrt{44}}{2}$ or $x = \dfrac{6 - \sqrt{44}}{2}$

≈ 6.3 ≈ -0.3

The solution set is $\{-0.3, 6.3\}$.

Exercises

Solve each equation by using the Quadratic Formula. Round to the nearest tenth if necessary.

1. $x^2 - 3x + 2 = 0$
2. $m^2 - 8m = -16$
3. $16r^2 - 8r = -1$

4. $x^2 + 5x = 6$
5. $3x^2 + 2x = 8$
6. $8x^2 - 8x - 5 = 0$

7. $-4c^2 + 19c = 21$
8. $2p^2 + 6p = 5$
9. $48x^2 + 22x - 15 = 0$

10. $8x^2 - 4x = 24$
11. $2p^2 + 5p = 8$
12. $8y^2 + 9y - 4 = 0$

13. $2x^2 + 9x + 4 = 0$
14. $8y^2 + 17y + 2 = 0$

15. $3z^2 + 5z - 2 = 0$
16. $-2x^2 + 8x + 4 = 0$

17. $a^2 + 3a = 2$
18. $2y^2 - 6y + 4 = 0$

NAME _____ DATE _____ PERIOD _____

9-4 Study Guide and Intervention (continued)

Solving Quadratic Equations by Using the Quadratic Formula

The Discriminant In the Quadratic Formula, $x = \dfrac{-b \pm \sqrt{b^2 - 4ac}}{2a}$, the expression under the radical sign, $b^2 - 4ac$, is called the **discriminant**. The discriminant can be used to determine the number of real roots for a quadratic equation.

Case 1: $b^2 - 4ac < 0$	Case 2: $b^2 - 4ac = 0$	Case 3: $b^2 - 4ac > 0$
no real roots	one real root	two real roots

Example
State the value of the discriminant for each equation. Then determine the number of real roots.

a. $12x^2 + 5x = 4$

Write the equation in standard form.

$12x^2 + 5x = 4$ Original equation
$12x^2 + 5x - 4 = 4 - 4$ Subtract 4 from each side.
$12x^2 + 5x - 4 = 0$ Simplify.

Now find the discriminant.
$b^2 - 4ac = (5)^2 - 4(12)(-4)$
$= 217$

Since the discriminant is positive, the equation has two real roots.

b. $2x^2 + 3x = -4$

$2x^2 + 3x = -4$ Original equation
$2x^2 + 3x + 4 = -4 + 4$ Add 4 to each side.
$2x^2 + 3x + 4 = 0$ Simplify.

$b^2 - 4ac = (3)^2 - 4(2)(4)$
$= -23$

Since the discriminant is negative, the equation has no real roots.

Exercises

State the value of the discriminant for each equation. Then determine the number of real roots of the equation.

1. $3x^2 + 2x - 3 = 0$

2. $3n^2 - 7n - 8 = 0$

3. $2d^2 - 10d - 9 = 0$

4. $4x^2 = x + 4$

5. $3x^2 - 13x = 10$

6. $6x^2 - 10x + 10 = 0$

7. $2k^2 - 20 = -k$

8. $6p^2 = -11p - 40$

9. $9 - 18x + 9x^2 = 0$

10. $12x^2 + 9 = -6x$

11. $9a^2 = 81$

12. $16y^2 + 16y + 4 = 0$

13. $8x^2 + 9x = 2$

14. $4a^2 - 4a + 4 = 3$

15. $3b^2 - 18b = -14$

9-5 Study Guide and Intervention
Exponential Functions

Graph Exponential Functions

Exponential Function	a function defined by an equation of the form $y = a^x$, where $a > 0$ and $a \neq 1$

You can use values of x to find ordered pairs that satisfy an exponential function. Then you can use the ordered pairs to graph the function.

Example 1 Graph $y = 3^x$. State the y-intercept.

x	y
−2	$\frac{1}{9}$
−1	$\frac{1}{3}$
0	1
1	3
2	9

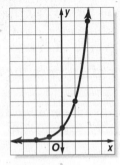

The y-intercept is 1.

Example 2 Graph $y = \left(\frac{1}{4}\right)^x$. Use the graph to determine the approximate value of $\left(\frac{1}{4}\right)^{-0.5}$.

x	y
−2	16
−1	4
0	1
1	$\frac{1}{4}$
2	$\frac{1}{16}$

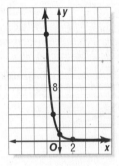

The value of $\left(\frac{1}{4}\right)^{-0.5}$ is about 2.

Exercises

1. Graph $y = 0.3^x$. State the y-intercept. Then use the graph to determine the approximate value of $0.3^{-1.5}$. Use a calculator to confirm the value.

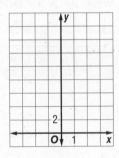

Graph each function. State the y-intercept.

2. $y = 3^x + 1$

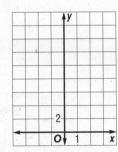

3. $y = \left(\frac{1}{3}\right)^x + 1$

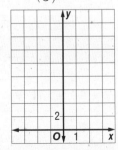

4. $y = \left(\frac{1}{2}\right)^x - 2$

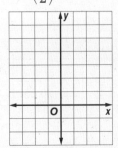

Study Guide and Intervention   Glencoe Algebra 1

NAME _____ DATE _____ PERIOD _____

9-5 Study Guide and Intervention *(continued)*
Exponential Functions

Identify Exponential Behavior It is sometimes useful to know if a set of data is exponential. One way to tell is to observe the shape of the graph. Another way is to observe the pattern in the set of data.

Example Determine whether the set of data displays exponential behavior.

x	0	2	4	6	8	10
y	64	32	16	8	4	2

Method 1: Graph the Data

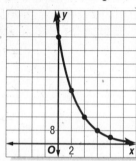

The graph shows rapidly decreasing values of *y* as *x* increases. This is characteristic of exponential behavior.

Method 2: Look for a Pattern

The domain values increase by regular intervals of 2, while the range values have a common factor of $\frac{1}{2}$. Since the domain values increase by regular intervals and the range values have a common factor, the data are probably exponential.

Exercises

Determine whether the data in each table display exponential behavior. Explain why or why not.

1.
x	0	1	2	3
y	5	10	15	20

2.
x	0	1	2	3
y	3	9	27	81

3.
x	−1	1	3	5
y	32	16	8	4

4.
x	−1	0	1	2	3
y	3	3	3	3	3

5.
x	−5	0	5	10
y	1	0.5	0.25	0.125

6.
x	0	1	2	3	4
y	$\frac{1}{3}$	$\frac{1}{9}$	$\frac{1}{27}$	$\frac{1}{81}$	$\frac{1}{243}$

NAME _____ DATE _____ PERIOD _____

9-6 Study Guide and Intervention

Growth and Decay

Exponential Growth Population increases and growth of monetary investments are examples of **exponential growth**. This means that an initial amount increases at a steady rate over time.

Exponential Growth	The general equation for exponential growth is $y = C(1 + r)^t$. • y represents the final amount. • C represents the initial amount. • r represents the rate of change expressed as a decimal. • t represents time.

Example 1 POPULATION The population of Johnson City in 2000 was 25,000. Since then, the population has grown at an average rate of 3.2% each year.

a. Write an equation to represent the population of Johnson City since 2000.

The rate 3.2% can be written as 0.032.

$y = C(1 + r)^t$
$y = 25{,}000(1 + 0.032)^t$
$y = 25{,}000(1.032)^t$

b. According to the equation, what will the population of Johnson City be in the year 2010?

In 2010, t will equal 2010 − 2000 or 10. Substitute 10 for t in the equation from part a.

$y = 25{,}000(1.032)^{10}$ $t = 10$
$\approx 34{,}256$

In 2010, the population of Johnson City will be about 34,256.

Example 2 INVESTMENT The Garcias have $12,000 in a savings account. The bank pays 3.5% interest on savings accounts, compounded monthly. Find the balance in 3 years.

The rate 3.5% can be written as 0.035.

The special equation for compound interest is $A = P\left(1 + \dfrac{r}{n}\right)^{nt}$, where A represents the balance, P is the initial amount, r represents the annual rate expressed as a decimal, n represents the number of times the interest is compounded each year, and t represents the number of years the money is invested.

$A = P\left(1 + \dfrac{r}{n}\right)^{nt}$

$A = 12{,}000\left(1 + \dfrac{0.035}{12}\right)^{36}$

$A \approx 12{,}000(1.00292)^{36}$

$A \approx 13{,}328.09$

In three years, the balance of the account will be $13,326.49.

Exercises

1. **POPULATION** The population of the United States has been increasing at an average annual rate of 0.91%. If the population of the United States was about 297,411,400 in the year 2005, predict the U. S. population in the year 2009. Source: U. S. Census Bureau

2. **INVESTMENT** Determine the amount of an investment of $2500 if it is invested at in interest rate of 5.25% compounded monthly for 4 years.

3. **POPULATION** It is estimated that the population of the world is increasing at an average annual rate of 1.3%. If the population of the world was about 6,472,416,997 in the year 2005, predict the world population in the year 2012. Source: U. S. Census Bureau

4. **INVESTMENT** Determine the amount of an investment of $100,000 if it is invested at an interest rate of 5.2% compounded quarterly for 12 years.

Study Guide and Intervention   Glencoe Algebra 1

9-6 Study Guide and Intervention (continued)
Growth and Decay

Exponential Decay Radioactive decay and depreciation are examples of **exponential decay**. This means that an initial amount decreases at a steady rate over a period of time.

Exponential Decay	The general equation for exponential decay is $y = C(1 - r)^t$. • y represents the final amount. • C represents the initial amount. • r represents the rate of decay expressed as a decimal. • t represents time.

Example **DEPRECIATION** The original price of a tractor was $45,000. The value of the tractor decreases at a steady rate of 12% per year.

a. Write an equation to represent the value of the tractor since it was purchased.

The rate 12% can be written as 0.12.
$y = C(1 - r)^t$ General equation for exponential decay
$y = 45,000(1 - 0.12)^t$ $C = 45,000$ and $r = 0.12$
$y = 45,000(0.88)^t$ Simplify.

b. What is the value of the tractor in 5 years?

$y = 45,000(0.88)^t$ Equation for decay from part a
$y = 45,000(0.88)^5$ $t = 5$
$y \approx 23,747.94$ Use a calculator.

In 5 years, the tractor will be worth about $23,747.94.

Exercises

1. **POPULATION** The population of Bulgaria has been decreasing at an annual rate of 0.89%. If the population of Bulgaria was about 7,450,349 in the year 2005, predict its population in the year 2015. **Source:** U. S. Census Bureau

2. **DEPRECIATION** Carl Gossell is a machinist. He bought some new machinery for about $125,000. He wants to calculate the value of the machinery over the next 10 years for tax purposes. If the machinery depreciates at the rate of 15% per year, what is the value of the machinery (to the nearest $100) at the end of 10 years?

3. **ARCHAEOLOGY** The *half-life* of a radioactive element is defined as the time that it takes for one-half a quantity of the element to decay. Radioactive Carbon-14 is found in all living organisms and has a half-life of 5730 years. Consider a living organism with an original concentration of Carbon-14 of 100 grams.

   a. If the organism lived 5730 years ago, what is the concentration of Carbon-14 today?

   b. If the organism lived 11,460 years ago, determine the concentration of Carbon-14 today.

4. **DEPRECIATION** A new car costs $32,000. It is expected to depreciate 12% each year for 4 years and then depreciate 8% each year thereafter. Find the value of the car in 6 years.

NAME _____ DATE _____ PERIOD _____

10-1 Study Guide and Intervention

Simplifying Radical Expressions

Product Property of Square Roots The **Product Property of Square Roots** and prime factorization can be used to simplify expressions involving irrational square roots. When you simplify radical expressions with variables, use absolute value to ensure nonnegative results.

Product Property of Square Roots	For any numbers a and b, where $a \geq 0$ and $b \geq 0$, $\sqrt{ab} = \sqrt{a} \cdot \sqrt{b}$.

Example 1 Simplify $\sqrt{180}$.

$\sqrt{180} = \sqrt{2 \cdot 2 \cdot 3 \cdot 3 \cdot 5}$ Prime factorization of 180
$= \sqrt{2^2} \cdot \sqrt{3^2} \cdot \sqrt{5}$ Product Property of Square Roots
$= 2 \cdot 3 \cdot \sqrt{5}$ Simplify.
$= 6\sqrt{5}$ Simplify.

Example 2 Simplify $\sqrt{120a^2 \cdot b^5 \cdot c^4}$.

$\sqrt{120a^2 \cdot b^5 \cdot c^4}$
$= \sqrt{2^3 \cdot 3 \cdot 5 \cdot a^2 \cdot b^5 \cdot c^4}$
$= \sqrt{2^2} \cdot \sqrt{2} \cdot \sqrt{3} \cdot \sqrt{5} \cdot \sqrt{a^2} \cdot \sqrt{b^4 \cdot b} \cdot \sqrt{c^4}$
$= 2 \cdot \sqrt{2} \cdot \sqrt{3} \cdot \sqrt{5} \cdot |a| \cdot b^2 \cdot \sqrt{b} \cdot c^2$
$= 2|a|b^2c^2\sqrt{30b}$

Exercises Simplify.

1. $\sqrt{28}$ 2. $\sqrt{68}$ 3. $\sqrt{60}$ 4. $\sqrt{75}$

5. $\sqrt{162}$ 6. $\sqrt{3} \cdot \sqrt{6}$ 7. $\sqrt{2} \cdot \sqrt{5}$ 8. $\sqrt{5} \cdot \sqrt{10}$

9. $\sqrt{4a^2}$ 10. $\sqrt{9x^4}$ 11. $\sqrt{300a^4}$ 12. $\sqrt{128c^6}$

13. $4\sqrt{10} \cdot 3\sqrt{6}$ 14. $\sqrt{3x^2} \cdot 3\sqrt{3x^4}$ 15. $\sqrt{20a^2b^4}$ 16. $\sqrt{100x^3y}$

17. $\sqrt{24a^4b^2}$ 18. $\sqrt{81x^4y^2}$ 19. $\sqrt{150a^2b^2c}$

20. $\sqrt{72a^6b^3c^2}$ 21. $\sqrt{45x^2y^5z^8}$ 22. $\sqrt{98x^4y^6z^2}$

Study Guide and Intervention   125   Glencoe Algebra 1

NAME _____ DATE _____ PERIOD _____

10-1 Study Guide and Intervention (continued)
Simplifying Radical Expressions

Quotient Property of Square Roots A fraction containing radicals is in simplest form if no radicals are left in the denominator. The **Quotient Property of Square Roots** and **rationalizing the denominator** can be used to simplify radical expressions that involve division. When you rationalize the denominator, you multiply the numerator and denominator by a radical expression that gives a rational number in the denominator.

Quotient Property of Square Roots	For any numbers a and b, where $a \geq 0$ and $b > 0$, $\sqrt{\dfrac{a}{b}} = \dfrac{\sqrt{a}}{\sqrt{b}}$.

Example Simplify $\sqrt{\dfrac{56}{45}}$.

$\sqrt{\dfrac{56}{45}} = \sqrt{\dfrac{4 \cdot 14}{9 \cdot 5}}$

$= \dfrac{2 \cdot \sqrt{14}}{3 \cdot \sqrt{5}}$ Simplify the numerator and denominator.

$= \dfrac{2\sqrt{14}}{3\sqrt{5}} \cdot \dfrac{\sqrt{5}}{\sqrt{5}}$ Multiply by $\dfrac{\sqrt{5}}{\sqrt{5}}$ to rationalize the denominator.

$= \dfrac{2\sqrt{70}}{15}$ Product Property of Square Roots

Exercises Simplify.

1. $\dfrac{\sqrt{9}}{\sqrt{18}}$

2. $\dfrac{\sqrt{8}}{\sqrt{24}}$

3. $\dfrac{\sqrt{100}}{\sqrt{121}}$

4. $\dfrac{\sqrt{75}}{\sqrt{3}}$

5. $\dfrac{8\sqrt{2}}{2\sqrt{8}}$

6. $\sqrt{\dfrac{2}{5}} \cdot \sqrt{\dfrac{6}{5}}$

7. $\sqrt{\dfrac{3}{4}} \cdot \sqrt{\dfrac{5}{2}}$

8. $\sqrt{\dfrac{5}{7}} \cdot \sqrt{\dfrac{2}{5}}$

9. $\sqrt{\dfrac{3a^2}{10b^6}}$

10. $\sqrt{\dfrac{x^6}{y^4}}$

11. $\sqrt{\dfrac{100a^4}{144b^8}}$

12. $\sqrt{\dfrac{75b^3c^6}{a^2}}$

13. $\dfrac{\sqrt{4}}{3 - \sqrt{5}}$

14. $\dfrac{\sqrt{8}}{2 + \sqrt{3}}$

15. $\dfrac{\sqrt{5}}{2 + \sqrt{5}}$

16. $\dfrac{\sqrt{8}}{2\sqrt{7} + 4\sqrt{10}}$

Study Guide and Intervention   Glencoe Algebra 1

10-2 Study Guide and Intervention

Operations with Radical Expressions

Add and Subtract Radical Expressions When adding or subtracting radical expressions, use the Associative and Distributive Properties to simplify the expressions. If radical expressions are not in simplest form, simplify them.

Example 1 Simplify $10\sqrt{6} - 5\sqrt{3} + 6\sqrt{3} - 4\sqrt{6}$.

$10\sqrt{6} - 5\sqrt{3} + 6\sqrt{3} - 4\sqrt{6} = (10 - 4)\sqrt{6} + (-5 + 6)\sqrt{3}$ Associative and Distributive Properties
$= 6\sqrt{6} + \sqrt{3}$ Simplify.

Example 2 Simplify $3\sqrt{12} + 5\sqrt{75}$.

$3\sqrt{12} + 5\sqrt{75} = 3\sqrt{2^2 \cdot 3} + 5\sqrt{5^2 \cdot 3}$ Simplify.
$= 3 \cdot 2\sqrt{3} + 5 \cdot 5\sqrt{3}$ Simplify.
$= 6\sqrt{3} + 25\sqrt{3}$ Simplify.
$= 31\sqrt{3}$ Distributive Property

Exercises Simplify.

1. $2\sqrt{5} + 4\sqrt{5}$

2. $\sqrt{6} - 4\sqrt{6}$

3. $\sqrt{8} - \sqrt{2}$

4. $3\sqrt{75} + 2\sqrt{5}$

5. $\sqrt{20} + 2\sqrt{5} - 3\sqrt{5}$

6. $2\sqrt{3} + \sqrt{6} - 5\sqrt{3}$

7. $\sqrt{12} + 2\sqrt{3} - 5\sqrt{3}$

8. $3\sqrt{6} + 3\sqrt{2} - \sqrt{50} + \sqrt{24}$

9. $\sqrt{8a} - \sqrt{2a} + 5\sqrt{2a}$

10. $\sqrt{54} + \sqrt{24}$

11. $\sqrt{3} + \sqrt{\dfrac{1}{3}}$

12. $\sqrt{12} + \sqrt{\dfrac{1}{3}}$

13. $\sqrt{54} - \sqrt{\dfrac{1}{6}}$

14. $\sqrt{80} - \sqrt{20} + \sqrt{180}$

15. $\sqrt{50} + \sqrt{18} - \sqrt{75} + \sqrt{27}$

16. $2\sqrt{3} - 4\sqrt{45} + 2\sqrt{\dfrac{1}{3}}$

17. $\sqrt{125} - 2\sqrt{\dfrac{1}{5}} + \sqrt{\dfrac{1}{3}}$

18. $\sqrt{\dfrac{2}{3}} + 3\sqrt{3} - 4\sqrt{\dfrac{1}{12}}$

Study Guide and Intervention

Glencoe Algebra 1

10-2 Study Guide and Intervention (continued)

Operations with Radical Expressions

Multiply Radical Expressions Multiplying two radical expressions with different radicands is similar to multiplying binomials.

Example Multiply $(3\sqrt{2} - 2\sqrt{5})(4\sqrt{20} + \sqrt{8})$.

Use the FOIL method.

$(3\sqrt{2} - 2\sqrt{5})(4\sqrt{20} + \sqrt{8}) = (3\sqrt{2})(4\sqrt{20}) + (3\sqrt{2})(\sqrt{8}) + (-2\sqrt{5})(4\sqrt{20}) + (-2\sqrt{5})(\sqrt{8})$

$= 12\sqrt{40} + 3\sqrt{16} - 8\sqrt{100} - 2\sqrt{40}$ Multiply.

$= 12\sqrt{2^2 \cdot 10} + 3 \cdot 4 - 8 \cdot 10 - 2\sqrt{2^2 \cdot 10}$ Simplify.

$= 24\sqrt{10} + 12 - 80 - 4\sqrt{10}$ Simplify.

$= 20\sqrt{10} - 68$ Combine like terms.

Exercises Find each product.

1. $2(\sqrt{3} + 4\sqrt{5})$

2. $\sqrt{6}(\sqrt{3} - 2\sqrt{6})$

3. $\sqrt{5}(\sqrt{5} - \sqrt{2})$

4. $\sqrt{2}(3\sqrt{7} + 2\sqrt{5})$

5. $(2 - 4\sqrt{2})(2 + 4\sqrt{2})$

6. $(3 + \sqrt{6})^2$

7. $(2 - 2\sqrt{5})^2$

8. $3\sqrt{2}(\sqrt{8} + \sqrt{24})$

9. $\sqrt{8}(\sqrt{2} + 5\sqrt{8})$

10. $(\sqrt{5} - 3\sqrt{2})(\sqrt{5} + 3\sqrt{2})$

11. $(\sqrt{3} + \sqrt{6})^2$

12. $(\sqrt{2} - 2\sqrt{3})^2$

13. $(\sqrt{5} - \sqrt{2})(\sqrt{2} + \sqrt{6})$

14. $(\sqrt{8} - \sqrt{2})(\sqrt{3} + \sqrt{6})$

15. $(\sqrt{5} - \sqrt{18})(7\sqrt{5} + \sqrt{3})$

16. $(2\sqrt{3} - \sqrt{45})(\sqrt{12} + 2\sqrt{6})$

17. $(2\sqrt{5} - 2\sqrt{3})(\sqrt{10} + \sqrt{6})$

18. $(\sqrt{2} + 3\sqrt{3})(\sqrt{12} - 4\sqrt{8})$

Study Guide and Intervention

Glencoe Algebra 1

NAME _____ DATE _____ PERIOD ____

10-3 Study Guide and Intervention

Radical Equations

Radical Equations Equations containing radicals with variables in the radicand are called **radical equations**. These can be solved by first using the following steps.

Step 1	Isolate the radical on one side of the equation.
Step 2	Square each side of the equation to eliminate the radical.

Example 1 Solve $16 = \dfrac{\sqrt{x}}{2}$ for x.

$16 = \dfrac{\sqrt{x}}{2}$ Original equation

$2(16) = 2\left(\dfrac{\sqrt{x}}{2}\right)$ Multiply each side by 2.

$32 = \sqrt{x}$ Simplify.

$(32)^2 = (\sqrt{x})^2$ Square each side.

$1024 = x$ Simplify.

The solution is 1024, which checks in the original equation.

Example 2 Solve $\sqrt{4x-7} + 2 = 7$.

$\sqrt{4x-7} + 2 = 7$ Original equation

$\sqrt{4x-7} + 2 - 2 = 7 - 2$ Subtract 2 from each side.

$\sqrt{4x-7} = 5$ Simplify.

$(\sqrt{4x-7})^2 = 5^2$ Square each side.

$4x - 7 = 25$ Simplify.

$4x - 7 + 7 = 25 + 7$ Add 7 to each side.

$4x = 32$ Simplify.

$x = 8$ Divide each side by 4.

The solution is 8, which checks in the original equation.

Exercises

Solve each equation. Check your solution.

1. $\sqrt{a} = 8$

2. $\sqrt{a} + 6 = 32$

3. $2\sqrt{x} = 8$

4. $7 = \sqrt{26-n}$

5. $\sqrt{-a} = 6$

6. $\sqrt{3r^2} = 3$

7. $2\sqrt{3} = \sqrt{y}$

8. $2\sqrt{3a} - 2 = 7$

9. $\sqrt{x-4} = 6$

10. $\sqrt{2c+3} = 5$

11. $\sqrt{3b-2} + 19 = 24$

12. $\sqrt{4x-1} = 3$

13. $\sqrt{3r+2} = 2\sqrt{3}$

14. $\sqrt{\dfrac{x}{2}} = \dfrac{1}{2}$

15. $\sqrt{\dfrac{x}{8}} = 4$

16. $\sqrt{6x^2 + 5x} = 2$

17. $\sqrt{\dfrac{x}{3}} + 6 = 8$

18. $2\sqrt{\dfrac{3x}{5}} + 3 = 11$

Study Guide and Intervention   Glencoe Algebra 1

NAME _____ DATE _____ PERIOD _____

10-3 Study Guide and Intervention (continued)

Radical Equations

Extraneous Solutions To solve a radical equation with a variable on both sides, you need to square each side of the equation. Squaring each side of an equation sometimes produces **extraneous solutions**, or solutions that are not solutions of the original equation. Therefore, it is very important that you check each solution.

Example Solve $\sqrt{x+3} = x - 3$.

$$\sqrt{x+3} = x - 3 \qquad \text{Original equation}$$
$$(\sqrt{x+3})^2 = (x-3)^2 \qquad \text{Square each side.}$$
$$x + 3 = x^2 - 6x + 9 \qquad \text{Simplify.}$$
$$0 = x^2 - 7x + 6 \qquad \text{Subtract } x \text{ and 3 from each side.}$$
$$0 = (x - 1)(x - 6) \qquad \text{Factor.}$$
$$x - 1 = 0 \quad \text{or} \quad x - 6 = 0 \qquad \text{Zero Product Property}$$
$$x = 1 \qquad \qquad x = 6 \qquad \text{Solve.}$$

CHECK
$\sqrt{x+3} = x - 3$ $\qquad$ $\sqrt{x+3} = x - 3$
$\sqrt{1+3} \stackrel{?}{=} 1 - 3$ $\qquad$ $\sqrt{6+3} \stackrel{?}{=} 6 - 3$
$\sqrt{4} \stackrel{?}{=} -2$ $\qquad\qquad$ $\sqrt{9} \stackrel{?}{=} 3$
$2 \neq -2$ $\qquad\qquad\qquad$ $3 = 3 \checkmark$

Since $x = 1$ does not satisfy the original equation, $x = 6$ is the only solution.

Exercises

Solve each equation. Check your solution.

1. $\sqrt{a} = a$

2. $\sqrt{a+6} = a$

3. $2\sqrt{x} = x$

4. $n = \sqrt{2 - n}$

5. $\sqrt{-a} = a$

6. $\sqrt{10 - 6k} + 3 = k$

7. $\sqrt{y - 1} = y - 1$

8. $\sqrt{3a - 2} = a$

9. $\sqrt{x + 2} = x$

10. $\sqrt{2c + 5} = c - 5$

11. $\sqrt{3b + 6} = b + 2$

12. $\sqrt{4x - 4} = x$

13. $r + \sqrt{2 - r} = 2$

14. $\sqrt{x^2 + 10x} = x + 4$

15. $-2\sqrt{\dfrac{x}{8}} = 15$

16. $\sqrt{6x^2 - 4x} = x + 2$

17. $\sqrt{2y^2 - 64} = y$

18. $\sqrt{3x^2 + 12x + 1} = x + 5$

NAME _____ DATE _____ PERIOD _____

10-4 Study Guide and Intervention

The Pythagorean Theorem

The Pythagorean Theorem The side opposite the right angle in a right triangle is called the **hypotenuse**. This side is always the longest side of a right triangle. The other two sides are called the **legs** of the triangle. To find the length of any side of a right triangle, given the lengths of the other two sides, you can use the **Pythagorean Theorem**.

| Pythagorean Theorem | If a and b are the measures of the legs of a right triangle and c is the measure of the hypotenuse, then $c^2 = a^2 + b^2$. | 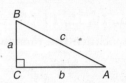 |

Example 1 Find the length of the hypotenuse of a right triangle if $a = 5$ and $b = 12$.

$c^2 = a^2 + b^2$ Pythagorean Theorem
$c^2 = 5^2 + 12^2$ $a = 5$ and $b = 12$
$c^2 = 169$ Simplify.
$c = \sqrt{169}$ Take the square root of each side.
$c = 13$

The length of the hypotenuse is 13.

Example 2 Find the length of a leg of a right triangle if $a = 8$ and $c = 10$.

$c^2 = a^2 + b^2$ Pythagorean Theorem
$10^2 = 8^2 + b^2$ $a = 8$ and $c = 10$
$100 = 64 + b^2$ Simplify.
$36 = b^2$ Subtract 64 from each side.
$b = \pm\sqrt{36}$ Take the square root of each side.
$b = \pm 6$

The length of the leg is 6.

Exercises

Find the length of each missing side. If necessary, round to the nearest hundredth.

1.

2.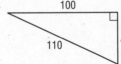

3.

If c is the measure of the hypotenuse of a right triangle, find each missing measure. If necessary, round to the nearest hundredth.

4. $a = 10, b = 12, c = ?$

5. $a = 9, b = 12, c = ?$

6. $a = 12, b = ?, c = 16$

7. $a = ?, b = 6, c = 8$

8. $a = ?, b = \sqrt{8}, c = \sqrt{18}$

9. $a = \sqrt{5}, b = \sqrt{10}, c = ?$

Study Guide and Intervention   Glencoe Algebra 1

10-4 Study Guide and Intervention (continued)
The Pythagorean Theorem

Right Triangles If a and b are the measures of the shorter sides of a triangle, c is the measure of the longest side, and $c^2 = a^2 + b^2$, then the triangle is a right triangle.

Example
Determine whether the following side measures form right triangles.

a. 10, 12, 14

Since the measure of the longest side is 14, let $c = 14$, $a = 10$, and $b = 12$.

$c^2 = a^2 + b^2$ Pythagorean Theorem
$14^2 \stackrel{?}{=} 10^2 + 12^2$ $a = 10, b = 12, c = 14$
$196 \stackrel{?}{=} 100 + 144$ Multiply.
$196 \neq 244$ Add.

Since $c^2 \neq a^2 + b^2$, the triangle is not a right triangle.

b. 7, 24, 25

Since the measure of the longest side is 25, let $c = 25$, $a = 7$, and $b = 24$.

$c^2 = a^2 + b^2$ Pythagorean Theorem
$25^2 \stackrel{?}{=} 7^2 + 24^2$ $a = 7, b = 24, c = 25$
$625 \stackrel{?}{=} 49 + 576$ Multiply.
$625 = 625$ Add.

Since $c^2 = a^2 + b^2$, the triangle is a right triangle.

Exercises
Determine whether the following side measures form right triangles.

1. 14, 48, 50

2. 6, 8, 10

3. 8, 8, 10

4. 90, 120, 150

5. 15, 20, 25

6. 4, 8, $4\sqrt{5}$

7. 2, 2, $\sqrt{8}$

8. 4, 4, $\sqrt{20}$

9. 25, 30, 35

10. 24, 36, 48

11. 18, 80, 82

12. 150, 200, 250

13. 100, 200, 300

14. 500, 1200, 1300

15. 700, 1000, 1300

NAME _____ DATE _____ PERIOD _____

10-5 Study Guide and Intervention
The Distance Formula

The Distance Formula The Pythagorean Theorem can be used to derive the **Distance Formula** shown below. The Distance Formula can then be used to find the distance between any two points in the coordinate plane.

Distance Formula	The distance between any two points with coordinates (x_1, y_1) and (x_2, y_2) is given by $d = \sqrt{(x_2 - x_1)^2 + (y_2 - y_1)^2}$.

Example 1 Find the distance between the points at $(-5, 2)$ and $(4, 5)$.

$d = \sqrt{(x_2 - x_1)^2 + (y_2 - y_1)^2}$ Distance Formula
$= \sqrt{(4 - (-5))^2 + (5 - 2)^2}$ $(x_1, y_1) = (-5, 2), (x_2, y_2) = (4, 5)$
$= \sqrt{9^2 + 3^2}$ Simplify.
$= \sqrt{81 + 9}$ Evaluate squares and simplify.
$= \sqrt{90}$

The distance is $\sqrt{90}$, or about 9.49 units.

Example 2 Jill draws a line segment from point $(1, 4)$ on her computer screen to point $(98, 49)$. How long is the segment?

$d = \sqrt{(x_2 - x_1)^2 + (y_2 - y_1)^2}$
$= \sqrt{(98 - 1)^2 + (49 - 4)^2}$
$= \sqrt{97^2 + 45^2}$
$= \sqrt{9409 + 2025}$
$= \sqrt{11,434}$

The segment is about 106.93 units long.

Exercises

Find the distance between each pair of points with the given coordinates. Express in simplest radical form and as decimal approximations rounded to the nearest hundredth if necessary.

1. $(1, 5), (3, 1)$
2. $(0, 0), (6, 8)$
3. $(-2, -8), (7, -3)$

4. $(6, -7), (-2, 8)$
5. $(1, 5), (-8, 4)$
6. $(3, -4), (-4, -4)$

7. $(-1, 4), (3, 2)$
8. $(0, 0), (-3, 5)$
9. $(2, -6), (-7, 1)$

10. $(-2, -5), (0, 8)$
11. $(3, 4), (0, 0)$
12. $(3, -4), (-4, -16)$

13. $(1, -1), (3, -2)$
14. $(-2, 0), (-3, -9)$
15. $(-9, 0), (-2, 5)$

16. $(2, -7), (-2, -2)$
17. $(1, -3), (-8, 21)$
18. $(-3, -5), (1, -8)$

10-5 Study Guide and Intervention (continued)

The Distance Formula

Find Coordinates If you know the coordinates of one point and only one coordinate of a second point, you can use the Distance Formula to find the missing coordinate of the second point.

Example Find the possible values of a if the distance between the points at $(-3, -2)$ and $(a, -5)$ is 5 units.

$d = \sqrt{(x_2 - x_1)^2 + (y_2 - y_1)^2}$ Distance Formula
$5 = \sqrt{(a - (-3))^2 + (-5 - (-2))^2}$ $(x_1, y_1) = (-3, -2), (x_2, y_2) = (a, -5)$, and $d = 5$
$5 = \sqrt{(a + 3)^2 + (-3)^2}$ Simplify.
$5 = \sqrt{a^2 + 6a + 9 + 9}$ Evaluate squares.
$5 = \sqrt{a^2 + 6a + 18}$ Simplify.
$5^2 = (\sqrt{a^2 + 6a + 18})^2$ Square each side.
$25 = a^2 + 6a + 18$ Simplify.
$0 = a^2 + 6a - 7$ Subtract 25 from each side.
$0 = (a + 7)(a - 1)$ Factor.
$a + 7 = 0$ or $a - 1 = 0$ Zero Product Property
$a = -7 \qquad\quad a = 1$

The value of a is -7 or 1.

Exercises

Find the possible values of a if the points with the given coordinates are the indicated distance apart.

1. $(1, a), (3, -2); d = \sqrt{5}$
2. $(0, 0), (a, 4); d = 5$
3. $(2, -1), (a, 3); d = 5$

4. $(1, -3), (a, 21); d = 25$
5. $(1, a), (-2, 4); d = 3$
6. $(3, -4), (-4, a); d = \sqrt{65}$

7. $(a, -4), (-3, -2); d = \sqrt{13}$
8. $(0, 3), (3, a); d = 3\sqrt{2}$
9. $(a, 3), (2, -4); d = \sqrt{74}$

10. $(-2, -5), (a, -2); d = 5$
11. $(3, 3), (-1, a); d = 5$
12. $(-1, -1), (4, a); d = \sqrt{41}$

13. $(a, 5), (-1, 2); d = \sqrt{45}$
14. $(4, -2), (a, 0); d = 2\sqrt{2}$
15. $(-2, 1), (a, -3); d = 2\sqrt{5}$

NAME _____ DATE _____ PERIOD _____

10-6 Study Guide and Intervention

Similar Triangles

Similar Triangles △RST is **similar** to △XYZ. The angles of the two triangles have equal measure. They are called **corresponding angles**. The sides opposite the corresponding angles are called **corresponding sides**.

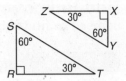

| Similar Triangles | If two triangles are similar, then the measures of their corresponding sides are proportional and the measures of their corresponding angles are equal. | △ABC ~ △DEF $\frac{AB}{DE} = \frac{BC}{EF} = \frac{AC}{DF}$ | 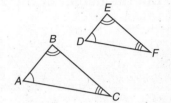 |

Example 1 Determine whether the pair of triangles is similar. Justify your answer.

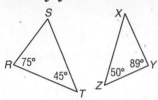

Since corresponding angles do not have the equal measures, the triangles are not similar.

Example 2 Determine whether the pair of triangles is similar. Justify your answer.

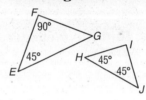

The measure of ∠G = 180° − (90° + 45°) = 45°.
The measure of ∠I = 180° − (45° + 45°) = 90°.
Since corresponding angles have equal measures, △EFG ~ △HIJ.

Exercises

Determine whether each pair of triangles is similar. Justify your answer.

1.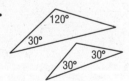

2.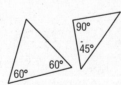

3.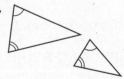

4.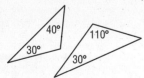

5.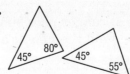

6.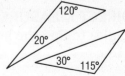

Study Guide and Intervention 135 Glencoe Algebra 1

10-6 Study Guide and Intervention (continued)

Similar Triangles

Find Unknown Measures If some of the measurements are known, proportions can be used to find the measures of the other sides of similar triangles.

Example INDIRECT MEASUREMENT
$\triangle ABC \sim \triangle AED$ in the figure at the right. Find the height of the apartment building.

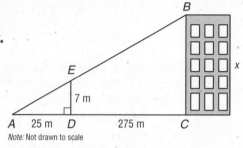

Let $BC = x$.

$\dfrac{ED}{BC} = \dfrac{AD}{AC}$

$\dfrac{7}{x} = \dfrac{25}{300}$ $ED = 7, AD = 25, AC = 300$

$25x = 2100$ Find the cross products.

$x = 84$

The apartment building is 84 meters high.

Exercises

For each set of measures, find the measures of the missing sides if $\triangle ABC \sim \triangle DEF$.

1. $c = 15, d = 8, e = 6, f = 10$

2. $c = 20, a = 12, b = 8, f = 15$

3. $a = 8, d = 8, e = 6, f = 7$

4. $a = 20, d = 10, e = 8, f = 10$

5. $c = 5, d = 10, e = 8, f = 8$

6. $a = 25, b = 20, c = 15, f = 12$

7. $b = 8, d = 8, e = 4, f = 10$

8. **INDIRECT MEASUREMENT** Bruce likes to amuse his brother by shining a flashlight on his hand and making a shadow on the wall. How far is it from the flashlight to the wall?

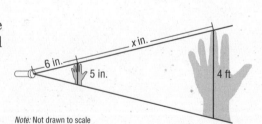

Note: Not drawn to scale

9. **INDIRECT MEASUREMENT** A forest ranger uses similar triangles to find the height of a tree. Find the height of the tree.

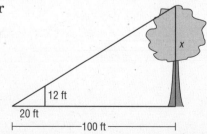

NAME _____ DATE _____ PERIOD _____

11-1 Study Guide and Intervention

Inverse Variation

Graph Inverse Variation Situations in which the values of y decrease as the values of x increase are examples of **inverse variation**. We say that y varies inversely as x, or y is inversely proportional to x.

Inverse Variation Equation	an equation of the form $xy = k$, where $k \neq 0$

Example 1 Suppose you drive 200 miles without stopping. The time it takes to travel a distance varies inversely as the rate at which you travel. Let x = speed in miles per hour and y = time in hours. Graph the variation.

The equation $xy = 200$ can be used to represent the situation. Use various speeds to make a table.

x	y
10	20
20	10
30	6.7
40	5
50	4
60	3.3

Example 2 Graph an inverse variation in which y varies inversely as x and $y = 3$ when $x = 12$.

Solve for k.
$xy = k$ Inverse variation equation
$12(3) = k$ $x = 12$ and $y = 3$
$36 = k$ Simplify.

Choose values for x and y, which have a product of 36.

x	y
−6	−6
−3	−12
−2	−18
2	18
3	12
6	6

Exercises

Graph each variation if y varies inversely as x.

1. $y = 9$ when $x = -3$

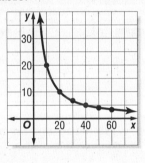

2. $y = 12$ when $x = 4$

3. $y = -25$ when $x = 5$

4. $y = 4$ when $x = 5$

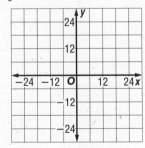

5. $y = -18$ when $x = -9$

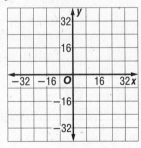

6. $y = 4.8$ when $x = 5.4$

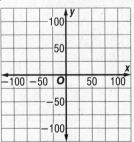

NAME _____ DATE _____ PERIOD _____

11-1 Study Guide and Intervention (continued)
Inverse Variation

Use Inverse Variation If two points (x_1, y_1) and (x_2, y_2) are solutions of an inverse variation, then $x_1 \cdot y_1 = k$ and $x_2 \cdot y_2 = k$.

Product Rule or Inverse Variation	the equation $x_1 \cdot y_1 = x_2 \cdot y_2$

From the product rule, you can form the proportion $\dfrac{x_1}{x_2} = \dfrac{y_2}{y_1}$.

Example
If y varies inversely as x and $y = 12$ when $x = 4$, find x when $y = 18$.

Method 1 Use the product rule.

$x_1 \cdot y_1 = x_2 \cdot y_2$ Product rule for inverse variation
$4 \cdot 12 = x_2 \cdot 18$ $x_1 = 4, y_1 = 12, y_2 = 18$
$\dfrac{48}{18} = x_2$ Divide each side by 18.
$\dfrac{8}{3} = x_2$ Simplify.

Method 2 Use a proportion.

$\dfrac{x_1}{x_2} = \dfrac{y_2}{y_1}$ Proportion for inverse variation
$\dfrac{4}{x_2} = \dfrac{18}{12}$ $x_1 = 4, y_1 = 12, y_2 = 18$
$48 = 18x_2$ Cross multiply.
$\dfrac{8}{3} = x_2$ Simplify.

Both methods show that $x_2 = \dfrac{8}{3}$ when $y = 18$.

Exercises

Write an inverse variation equation that relates x and y. Assume that y varies inversely as x. Then solve.

1. If $y = 10$ when $x = 5$, find y when $x = 2$.

2. If $y = 8$ when $x = -2$, find y when $x = 4$.

3. If $y = 70$ when $x = 7$, find y when $x = -5$.

4. If $y = 1.5$ when $x = 0.5$, find x when $y = 3$.

5. If $y = 100$ when $x = 120$, find x when $y = 20$.

6. If $y = -16$ when $x = 4$, find x when $y = 32$.

7. If $y = -7.5$ when $x = 25$, find y when $x = 5$.

8. If $y = \dfrac{1}{2}$ when $x = \dfrac{1}{8}$, find y when $x = \dfrac{1}{2}$.

9. If $y = \dfrac{3}{4}$ when $x = \dfrac{5}{4}$, find y when $x = \dfrac{1}{2}$.

10. **DRIVING** The Gerardi family can travel to Oshkosh, Wisconsin, from Chicago, Illinois, in 4 hours if they drive an average of 45 miles per hour. How long would it take them if they increased their average speed to 50 miles per hour?

11. **GEOMETRY** For a rectangle with given area, the width of the rectangle varies inversely as the length. If the width of the rectangle is 40 meters when the length is 5 meters, find the width of the rectangle when the length is 20 meters.

NAME _____ DATE _____ PERIOD _____

11-2 Study Guide and Intervention

Rational Expressions

Excluded Values of Rational Expressions

| Rational Expression | an algebraic fraction with numerator and denominator that are polynomials | Example: $\dfrac{x^2 + 1}{y^2}$ |

Because a rational expression involves division, the denominator cannot equal zero. Any value of the denominator that results in division by zero is called an **excluded value** of the denominator.

Example 1 State the excluded value of $\dfrac{4m - 8}{m + 2}$.

Exclude the values for which $m + 2 = 0$.
$m + 2 = 0$ The denominator cannot equal 0.
$m + 2 - 2 = 0 - 2$ Subtract 2 from each side.
$m = -2$ Simplify.
Therefore, m cannot equal -2.

Example 2 State the excluded values of $\dfrac{x^2 + 1}{x^2 - 9}$.

Exclude the values for which $x^2 - 9 = 0$.
$x^2 - 9 = 0$ The denominator cannot equal 0.
$(x + 3)(x - 3) = 0$ Factor.
$x + 3 = 0$ or $x - 3 = 0$ Zero Product Property
$\quad = -3 \qquad \qquad = 3$
Therefore, x cannot equal -3 or 3.

Exercises

State the excluded values for each rational expression.

1. $\dfrac{2b}{b - 8}$

2. $\dfrac{12 - a}{32 + a}$

3. $\dfrac{x^2 - 2}{x + 4}$

4. $\dfrac{m^2 - 4}{2m + 8}$

5. $\dfrac{2n - 12}{n^2 - 4}$

6. $\dfrac{2x + 18}{x^2 - 16}$

7. $\dfrac{x^2 - 4}{x^2 + 4x + 4}$

8. $\dfrac{a - 1}{a^2 + 5a + 6}$

9. $\dfrac{k^2 - 2k + 1}{k^2 + 4k + 3}$

10. $\dfrac{m^2 - 1}{2m^2 - m - 1}$

11. $\dfrac{25 - n^2}{n^2 - 4n - 5}$

12. $\dfrac{2x^2 + 5x + 1}{x^2 - 10x + 16}$

13. $\dfrac{n^2 - 2n - 3}{n^2 + 4n - 5}$

14. $\dfrac{y^2 - y - 2}{3y^2 - 12}$

15. $\dfrac{k^2 + 2k - 3}{k^2 - 20k + 64}$

16. $\dfrac{x^2 + 4x + 4}{4x^2 + 11x - 3}$

Study Guide and Intervention   Glencoe Algebra 1

NAME _____ DATE _____ PERIOD _____

11-2 Study Guide and Intervention (continued)

Rational Expressions

Simplify Rational Expressions Factoring polynomials is a useful tool for simplifying rational expressions. To simplify a rational expression, first factor the numerator and denominator. Then divide each by the greatest common factor.

Example 1 Simplify $\dfrac{54z^3}{24yz}$.

$\dfrac{54z^3}{24yz} = \dfrac{(6z)(9z^2)}{(6z)(4y)}$ The GCF of the numerator and the denominator is 6z.

$= \dfrac{\cancel{(6z)}(9z^2)}{\cancel{(6z)}(4y)}$ Divide the numerator and denominator by 6z.

$= \dfrac{9z^2}{4y}$ Simplify.

Example 2 Simplify $\dfrac{3x - 9}{x^2 - 5x + 6}$. State the excluded values of x.

$\dfrac{3x - 9}{x^2 - 5x + 6} = \dfrac{3(x - 3)}{(x - 2)(x - 3)}$ Factor.

$= \dfrac{3\cancel{(x-3)}}{(x-2)\cancel{(x-3)}}$ Divide by the GCF, $x - 3$.

$= \dfrac{3}{x - 2}$ Simplify.

Exclude the values for which $x^2 - 5x + 6 = 0$.
$x^2 - 5x + 6 = 0$
$(x - 2)(x - 3) = 0$
$x = 2$ or $x = 3$
Therefore, $x \neq 2$ and $x \neq 3$.

Exercises

Simplify each expression. State the excluded values of the variables.

1. $\dfrac{12ab}{a^2b^2}$

2. $\dfrac{7n^3}{21n^8}$

3. $\dfrac{x + 2}{x^2 - 4}$

4. $\dfrac{m^2 - 4}{m^2 + 6m + 8}$

5. $\dfrac{2n - 8}{n^2 - 16}$

6. $\dfrac{x^2 + 2x + 1}{x^2 - 1}$

7. $\dfrac{x^2 - 4}{x^2 + 4x + 4}$

8. $\dfrac{a^2 + 3a + 2}{a^2 + 5a + 6}$

9. $\dfrac{k^2 - 1}{k^2 + 4k + 3}$

10. $\dfrac{m^2 - 2m + 1}{2m^2 - m - 1}$

11. $\dfrac{n^2 - 25}{n^2 - 4n - 5}$

12. $\dfrac{x^2 + x - 6}{2x^2 - 8}$

13. $\dfrac{n^2 + 7n + 12}{n^2 + 2n - 8}$

14. $\dfrac{y^2 - y - 2}{y^2 - 10y + 16}$

NAME _____ DATE _____ PERIOD _____

11-3 Study Guide and Intervention

Multiplying Rational Expressions

Multiply Rational Expressions To multiply rational expressions, you multiply the numerators and multiply the denominators. Then simplify.

Example 1 Find $\dfrac{2c^2d}{5ab^2} \cdot \dfrac{a^2b}{3cd}$.

$\dfrac{2c^2d}{5ab^2} \cdot \dfrac{a^2b}{3cd} = \dfrac{2a^2bc^2d}{15ab^2cd}$ Multiply.

$= \dfrac{{}^1(abcd)(2ac)}{{}_1(abcd)(15b)}$ Simplify.

$= \dfrac{2ac}{15b}$ Simplify.

Example 2 Find $\dfrac{x^2 - 16}{2x + 8} \cdot \dfrac{x + 4}{x^2 + 8x + 16}$.

$\dfrac{x^2 - 16}{2x + 8} \cdot \dfrac{x + 4}{x^2 + 8x + 16} = \dfrac{(x-4)(x+4)}{2(x+4)} \cdot \dfrac{x+4}{(x+4)(x+4)}$ Factor.

$= \dfrac{(x-4)(x+4)^1}{2(x+4)_1} \cdot \dfrac{x+4\,^1}{(x+4)(x+4)_1}$ Simplify.

$= \dfrac{x - 4}{2x + 8}$ Multiply.

Exercises

Find each product.

1. $\dfrac{6ab}{a^2b^2} \cdot \dfrac{a^2}{b^2}$

2. $\dfrac{mn^2}{3} \cdot \dfrac{4}{mn}$

3. $\dfrac{x+2}{x-4} \cdot \dfrac{x-4}{x-1}$

4. $\dfrac{m-5}{8} \cdot \dfrac{16}{m-5}$

5. $\dfrac{2n-8}{n+2} \cdot \dfrac{2n+4}{n-4}$

6. $\dfrac{x^2 - 64}{2x + 16} \cdot \dfrac{x+8}{x^2 + 16x + 64}$

7. $\dfrac{8x + 8}{x^2 - 2x + 1} \cdot \dfrac{x-1}{2x+2}$

8. $\dfrac{a^2 - 25}{a + 2} \cdot \dfrac{a^2 - 4}{a - 5}$

9. $\dfrac{x^2 + 6x + 8}{2x^2 + 9x + 4} \cdot \dfrac{2x^2 - x - 1}{x^2 - 3x + 2}$

10. $\dfrac{m^2 - 1}{2m^2 - m - 1} \cdot \dfrac{2m + 1}{m^2 - 2m + 1}$

11. $\dfrac{n^2 - 1}{n^2 - 7n + 10} \cdot \dfrac{n^2 - 25}{n^2 + 6n + 5}$

12. $\dfrac{3p - 3q}{10pq} \cdot \dfrac{20p^2q^2}{p^2 - q^2}$

13. $\dfrac{a^2 + 7a + 12}{a^2 + 2a - 8} \cdot \dfrac{a^2 + 3a - 10}{a^2 + 2a - 8}$

14. $\dfrac{v^2 - 4v - 21}{3v^2 + 6v} \cdot \dfrac{v^2 + 8v}{v^2 + 11v + 24}$

11-3 Study Guide and Intervention (continued)

Multiplying Rational Expressions

Dimensional Analysis Multiplying fractions that involve units of measure can be simplified by dividing units of measure. This is similar to simplifying rational expressions by dividing out common factors.

Example The tank of a car holds 15 gallons of gasoline. You want to travel 1200 miles on a vacation. If the car averages 24 miles per gallon, how many tanks of gasoline can you expect to use for this trip?

$\dfrac{1200 \text{ miles}}{1 \text{ trip}} \cdot \dfrac{1 \text{ gallon}}{24 \text{ miles}} \cdot \dfrac{1 \text{ tank}}{15 \text{ gallons}}$ Use dimensional analysis.

$= \dfrac{1200 \cancel{\text{ miles}}}{1 \text{ trip}} \cdot \dfrac{1 \cancel{\text{ gallon}}}{24 \cancel{\text{ miles}}} \cdot \dfrac{1 \text{ tank}}{15 \cancel{\text{ gallons}}}$ Divide similar units.

$= \dfrac{1200}{360} \cdot \dfrac{\text{tank}}{\text{trip}}$ Simplify.

$= 3\dfrac{1}{3}$ tanks/trip   Simplify.

You can expect to use $3\dfrac{1}{3}$ tanks of gasoline for the trip.

Exercises

Find each product.

1. $\dfrac{1 \text{ mile}}{8 \text{ blocks}} \cdot \dfrac{32 \text{ blocks}}{3 \text{ hours}}$

2. $\dfrac{8 \text{ centimeters}}{1 \text{ pin}} \cdot \dfrac{1 \text{ meter}}{100 \text{ centimeters}} \cdot \dfrac{1 \text{ kilometer}}{1000 \text{ meters}}$

3. $\dfrac{50 \text{ miles}}{1 \text{ hour}} \cdot \dfrac{5280 \text{ feet}}{1 \text{ mile}} \cdot \dfrac{1 \text{ hour}}{60 \text{ minutes}} \cdot \dfrac{1 \text{ minute}}{60 \text{ seconds}}$

4. $\dfrac{2000 \text{ revolutions}}{1 \text{ minute}} \cdot \dfrac{60 \text{ minutes}}{1 \text{ hour}} \cdot \dfrac{24 \text{ hours}}{1 \text{ day}} \cdot \dfrac{7 \text{ days}}{1 \text{ week}}$

5. **CARPETING** Sarina's living room is 15 feet by 18 feet. What will it cost to carpet the living room if carpet costs $24 per square yard?

6. **CHARITY WALK** Suppose you participate in a walk for charity by asking for pledges. You walk at a constant rate of 2 miles per hour and have pledges totaling $15 per mile. How much do you earn for the charity if you walk for 16 hours?

7. **PAINTING** Suppose you estimate that the exterior of your house will need 15 gallons of paint. You use a special paint sprayer that holds one quart of paint. How many times do you need to fill the sprayer to paint the house?

8. **CALORIES** The number of calories used to play volleyball depends on your weight and how long you play. Playing volleyball expends 2.2 Calories per hour per pound of weight. If you weigh 130 pounds, how many Calories do you use in 1.5 hours?

9. **GARDENING** The rectangular lawns in a tract housing development each measure 100 feet by 200 feet. A gardener can cut 1000 square yards of grass per hour. How many of these lawns can he cut in $6\dfrac{2}{3}$ hours?

NAME _____ DATE _____ PERIOD _____

11-4 Study Guide and Intervention

Dividing Rational Expressions

Divide Rational Expressions To divide rational expressions, multiply by the reciprocal of the divisor. Then simplify.

Example 1 Find $\dfrac{12c^2d}{5a^2b^2} \div \dfrac{c^2d^2}{10ab}$.

$$\dfrac{12c^2d}{5a^2b^2} \div \dfrac{c^2d^2}{10ab} = \dfrac{12c^2d}{5a^2b^2} \times \dfrac{10ab}{c^2d^2}$$

$$= \dfrac{\cancel{12}^{12}\cancel{c^2}^{1}\cancel{d}^{1}}{\cancel{5}_{1}\cancel{a^2}_{a}\cancel{b^2}_{b}} \times \dfrac{\cancel{10}^{2}\cancel{a}\cancel{b}^{1}}{\cancel{c^2}_{1}\cancel{d^2}_{d}}$$

$$= \dfrac{24}{abd}$$

Example 2 Find $\dfrac{x^2 + 6x - 27}{x^2 + 11x + 18} \div \dfrac{x - 3}{x^2 + x - 2}$.

$$\dfrac{x^2 + 6x - 27}{x^2 + 11x + 18} \div \dfrac{x - 3}{x^2 + x - 2} = \dfrac{x^2 + 6x - 27}{x^2 + 11x + 18} \times \dfrac{x^2 + x - 2}{x - 3}$$

$$= \dfrac{(x+9)(x-3)}{(x+9)(x+2)} \times \dfrac{(x+2)(x-1)}{x-3}$$

$$= \dfrac{{}^1\cancel{(x+9)}\cancel{(x-3)}^1}{{}_1\cancel{(x+9)}\cancel{(x+2)}_1} \times \dfrac{{}^1\cancel{(x+2)}(x-1)}{\cancel{x-3}_1}$$

$$= x - 1$$

Exercises

Find each quotient.

1. $\dfrac{12ab}{a^2b^2} \div \dfrac{b}{a}$

2. $\dfrac{n}{4} \div \dfrac{n}{m}$

3. $\dfrac{3xy^2}{8} \div 6xy$

4. $\dfrac{m-5}{8} \div \dfrac{m-5}{16}$

5. $\dfrac{2n-4}{2n} \div \dfrac{n^2-4}{n}$

6. $\dfrac{y^2-36}{y^2-49} \div \dfrac{y+6}{y+7}$

7. $\dfrac{x^2-5x+6}{5} \div \dfrac{x-3}{15}$

8. $\dfrac{a^2b^3c}{3s^2t} \div \dfrac{6a^2bc}{8st^2u}$

9. $\dfrac{x^2+6x+8}{x^2+4x+4} \div \dfrac{x+4}{x+2}$

10. $\dfrac{m^2-49}{m} \div \dfrac{m^2-13m+42}{3m^2}$

11. $\dfrac{n^2-5n+6}{n^2+3n} \div \dfrac{3-n}{4n+12}$

12. $\dfrac{p^2-2pq+q^2}{p+q} \div \dfrac{p^2-q^2}{p+q}$

13. $\dfrac{a^2+7a+12}{a^2+3a-10} \div \dfrac{a^2-9}{a^2-25}$

14. $\dfrac{a^2-9}{2a^2+13a-7} \div \dfrac{a+3}{4a^2-1}$

Study Guide and Intervention 143 Glencoe Algebra 1

NAME _____ DATE _____ PERIOD _____

11-4 Study Guide and Intervention (continued)

Dividing Rational Expressions

Dimensional Analysis Dividing rational expressions that involve units of measure can also be simplified by dimensional analysis.

Example **SKATING** A speed skater skated 1000 meters in 1.14 minutes. Find his speed in meters per minute.

Use the formula for time and distance, rate · time = distance.

$r \cdot t = d$
$r \cdot 1.14 \text{ minutes} = 1000 \text{ meters}$ $t = 1.14$ minutes, $d = 1000$ meters
$\phantom{r \cdot 1.14 \text{ minutes}} r = 1000 \text{ meters} \div 1.14 \text{ minutes}$ Divide each side by 1.14 minutes.
$\phantom{r \cdot 1.14 \text{ minutes} r} = 1000 \text{ m} \cdot \dfrac{1}{1.14 \text{ minutes}}$ Multiply by the reciprocal.
$\phantom{r \cdot 1.14 \text{ minutes} r} = \dfrac{1000 \text{ meters}}{1.14 \text{ minutes}}$ Multiply.
$\phantom{r \cdot 1.14 \text{ minutes} r} \approx 877.2 \text{ meters/minute}$ Express as a unit rate.

Exercises

Complete.

1. $32 \text{ yd}^3 = \underline{} \text{ ft}^3$

2. $0.48 \text{ m}^3 = \underline{} \text{ cm}^3$

3. $144 \text{ in}^3 = \underline{} \text{ ft}^3$

4. $1400 \text{ m/min} = \underline{} \text{ km/min}$

5. $48 \text{ plants/ft}^2 = \underline{} \text{ plants/yd}^2$

6. $\$9/\text{h} = \$\underline{}/\text{min}$

7. $40 \text{ m/s} = \underline{} \text{ m/min}$

8. $\$5.04/\text{doz} = \$\underline{}/\text{piece}$

9. Express 90 kilometers per hour in meters per second.

10. Express 55 miles per hour in miles per minute.

11. Express $4.48 per pound in dollars per ounce.

12. **FOOD PRICES** A 14-ounce box of cereal costs $3.92. If a 16-ounce box of the same cereal costs the same per ounce, how much does the 16-ounce box cost?

13. **BICYCLING** The tires on a bicycle are 30 inches in diameter.

   a. Find the number of revolutions per mile.

   b. How many times per minute do the tires revolve when the bicycle is traveling at 12 miles per hour? Round to the nearest whole number.

14. **COOKING** A certain recipe calls for $\dfrac{3}{4}$ cup sugar and it makes 36 cookies.

   a. If 8 tablespoons = $\dfrac{1}{2}$ cup, how many tablespoons of sugar is in each cookie?

   b. Each tablespoon of sugar contains 48 Calories. If each cookie has 160 Calories, what percent of the calories are from sugar?

NAME _____ DATE _____ PERIOD _____

11-5 Study Guide and Intervention

Dividing Polynomials

Divide Polynomials by Monomials To divide a polynomial by a monomial, divide each term of the polynomial by the monomial.

Example 1 Find $(4r^2 - 12r) \div (2r)$.

$$(4r^2 - 12r) \div 2r = \frac{4r^2 - 12r}{2r}$$
$$= \frac{4r^2}{2r} - \frac{12r}{2r} \quad \text{Divide each term.}$$
$$= \frac{\cancel{4r^2}}{\cancel{2r}} - \frac{\cancel{12r}^6}{\cancel{2r}_1} \quad \text{Simplify.}$$
$$= 2r - 6 \quad \text{Simplify.}$$

Example 2 Find $(3x^2 - 8x + 4) \div (4x)$.

$$(3x^2 - 8x + 4) \div 4x = \frac{3x^2 - 8x + 4}{4x}$$
$$= \frac{3x^2}{4x} - \frac{8x}{4x} + \frac{4}{4x}$$
$$= \frac{\cancel{3x^2}^{3x}}{\cancel{4x}_4} - \frac{8x}{4x} + \frac{4}{4x}$$
$$= \frac{3x}{4} - 2 + \frac{1}{x}$$

Exercises

Find each quotient.

1. $(x^3 + 2x^2 - x) \div x$

2. $(2x^3 + 12x^2 - 8x) \div (2x)$

3. $(x^2 + 3x - 4) \div x$

4. $(4m^2 + 6m - 8) \div (2m^2)$

5. $(3x^3 + 15x^2 - 21x) \div (3x)$

6. $(8m^2n^2 + 4mn - 8n) \div n$

7. $(8y^4 + 16y^2 - 4) \div (4y^2)$

8. $(16x^4y^2 + 24xy + 5) \div (xy)$

9. $\dfrac{15x^2 - 25x + 30}{5}$

10. $\dfrac{10a^2b + 12ab - 8b}{2a}$

11. $\dfrac{6x^3 + 9x^2 + 9}{3x}$

12. $\dfrac{m^2 - 12m + 42}{3m^2}$

13. $\dfrac{m^2n^2 - 5mn + 6}{m^2n^2}$

14. $\dfrac{p^2 - 4pq + 6q^2}{pq}$

15. $\dfrac{6a^2b^2 - 8ab + 12}{2a^2}$

16. $\dfrac{2x^2y^3 - 4x^2y^2 - 8xy}{2xy}$

17. $\dfrac{9x^2y^2z - 2xyz + 12x}{xy}$

18. $\dfrac{2a^3b^3 + 8a^2b^2 - 10ab + 12}{2a^2b^2}$

11-5 Study Guide and Intervention (continued)
Dividing Polynomials

Divide Polynomials by Binomials To divide a polynomial by a binomial, factor the dividend if possible and divide both dividend and divisor by the GCF. If the polynomial cannot be factored, use long division.

Example Find $(x^2 + 7x + 10) \div (x + 3)$.

Step 1 Divide the first term of the dividend, x^2 by the first term of the divisor, x.

$$\begin{array}{r} x \\ x+3 \overline{) x^2 + 7x + 10} \\ (-)\underline{x^2 + 3x} \\ 4x \end{array}$$

Multiply x and $x + 3$.
Subtract.

Step 2 Bring down the next term, 10. Divide the first term of $4x + 10$ by x.

$$\begin{array}{r} x + 4 \\ x+3 \overline{) x^2 + 7x + 10} \\ \underline{x^2 + 3x} \\ 4x + 10 \\ (-)\underline{4x + 12} \\ -2 \end{array}$$

Multiply 4 and $x + 3$.
Subtract.

The quotient is $x + 4$ with remainder -2. The quotient can be written as $x + 4 + \dfrac{-2}{x+3}$.

Exercises

Find each quotient.

1. $(b^2 - 5b + 6) \div (b - 2)$

2. $(x^2 - x - 6) \div (x - 3)$

3. $(x^2 + 3x - 4) \div (x - 1)$

4. $(m^2 + 2m - 8) \div (m + 4)$

5. $(x^2 + 5x + 6) \div (x + 2)$

6. $(m^2 + 4m + 4) \div (m + 2)$

7. $(2y^2 + 5y + 2) \div (y + 2)$

8. $(8y^2 - 15y - 2) \div (y - 2)$

9. $\dfrac{8x^2 - 6x - 9}{4x + 3}$

10. $\dfrac{m^2 - 5m - 6}{m - 6}$

11. $\dfrac{x^3 + 1}{x - 2}$

12. $\dfrac{6m^3 + 11m^2 + 4m + 35}{2m + 5}$

13. $\dfrac{6a^2 + 7a + 5}{2a + 5}$

14. $\dfrac{8p^3 + 27}{2p + 3}$

11-6 Study Guide and Intervention

Rational Expressions with Like Denominators

Add Rational Expressions To add rational expressions with like denominators, add the numerators and then write the sum over the common denominator. If possible, simplify the resulting rational expression.

Example 1
Find $\dfrac{5n}{15} + \dfrac{7n}{15}$.

$\dfrac{5n}{15} + \dfrac{7n}{15} = \dfrac{5n + 7n}{15}$ Add the numerators.

$= \dfrac{12n}{15}$ Simplify.

$= \dfrac{\cancel{12n}^{\,4n}}{\cancel{15}_{\,5}}$ Divide by 3.

$= \dfrac{4n}{5}$ Simplify.

Example 2
Find $\dfrac{3x}{x+2} + \dfrac{6}{x+2}$.

$\dfrac{3x}{x+2} + \dfrac{6}{x+2} = \dfrac{3x + 6}{x+2}$

$= \dfrac{3(x+2)}{x+2}$

$= \dfrac{3\cancel{(x+2)}^{\,1}}{\cancel{x+2}_{\,1}}$

$= \dfrac{3}{1}$ or 3

Exercises

Find each sum.

1. $\dfrac{3}{a} + \dfrac{4}{a}$

2. $\dfrac{x^2}{8} + \dfrac{x}{8}$

3. $\dfrac{x+3}{6} + \dfrac{x-2}{6}$

4. $\dfrac{m-8}{2} + \dfrac{m+4}{2}$

5. $\dfrac{2x}{x+5} + \dfrac{3x}{x+5}$

6. $\dfrac{m+4}{m-1} + \dfrac{m+4}{m-1}$

7. $\dfrac{y+5}{y+6} + \dfrac{1}{y+6}$

8. $\dfrac{3x+5}{5} + \dfrac{2x+10}{5}$

9. $\dfrac{2a-4}{a-4} + \dfrac{-a}{a-4}$

10. $\dfrac{m+1}{2m-1} + \dfrac{3m-3}{2m-1}$

11. $\dfrac{x+1}{x-2} + \dfrac{x-5}{x-2}$

12. $\dfrac{5a}{3b^2} + \dfrac{10a}{3b^2}$

13. $\dfrac{3x+2}{x+2} + \dfrac{x+6}{x+2}$

14. $\dfrac{a-4}{a+1} + \dfrac{a+6}{a+1}$

15. $\dfrac{2x+3}{x+3} + \dfrac{x+6}{x+3}$

16. $\dfrac{3a^2+4a}{a} + \dfrac{6a^2}{a}$

17. $\dfrac{-8x}{x-4} + \dfrac{4x+x^2}{x-4}$

18. $\dfrac{9a-14}{2a+1} + \dfrac{8a+16}{2a+1}$

NAME _____ DATE _____ PERIOD _____

11-6 Study Guide and Intervention (continued)

Rational Expressions with Like Denominators

Subtract Rational Expressions To subtract fractions with like denominators, subtract the numerators and then write the difference over the common denominator. If possible, simplify the resulting rational expression.

Example Find $\dfrac{3x+2}{x-2} - \dfrac{4x}{x-2}$.

$$\dfrac{3x+2}{x-2} - \dfrac{4x}{x-2} = \dfrac{3x+2-4x}{x-2} \quad \text{The common denominator is } x-2.$$

$$= \dfrac{2-x}{x-2} \quad \text{Subtract.}$$

$$= \dfrac{-1(x-2)}{x-2} \quad 2-x = -1(x-2)$$

$$= \dfrac{-1\cancel{(x-2)}^{1}}{\cancel{x-2}\,^{1}}$$

$$= \dfrac{-1}{1} \quad \text{Simplify.}$$

$$= -1$$

Exercises

Find each difference.

1. $\dfrac{3}{a} - \dfrac{5}{a}$

2. $\dfrac{5x}{8} - \dfrac{x}{8}$

3. $\dfrac{5x}{9} - \dfrac{x}{9}$

4. $\dfrac{11x}{15y} - \dfrac{x}{15y}$

5. $\dfrac{8t}{w+6} - \dfrac{3t}{w+6}$

6. $\dfrac{7m+1}{3m+1} - \dfrac{4m}{3m+1}$

7. $\dfrac{y+7}{y+6} - \dfrac{1}{y+6}$

8. $\dfrac{3y+5}{5} - \dfrac{2y}{5}$

9. $\dfrac{2a+8}{a+4} - \dfrac{a+4}{a+4}$

10. $\dfrac{m+1}{2m-3} - \dfrac{3m-2}{2m-3}$

11. $\dfrac{x^2+x}{x} - \dfrac{x^2+5x}{x}$

12. $\dfrac{5a+2}{a^2} - \dfrac{4a+2}{a^2}$

13. $\dfrac{c^2}{c+2} - \dfrac{4}{c+2}$

14. $\dfrac{a-4}{a+1} - \dfrac{a+6}{a+1}$

15. $\dfrac{x^2+2x}{x-4} - \dfrac{x^2+8}{x-4}$

16. $\dfrac{a}{a^2-1} - \dfrac{1}{a^2-1}$

17. $\dfrac{4x-4y}{4x+4y} - \dfrac{4x}{4x+4y}$

18. $\dfrac{x^2+x}{x-2} - \dfrac{6}{x-2}$

NAME _____ DATE _____ PERIOD _____

11-7 Study Guide and Intervention

Rational Expressions with Unlike Denominators

Add Rational Expressions Adding rational expressions with unlike denominators is similar to adding fractions with unlike denominators.

Adding Rational Expressions	**Step 1** Find the LCD of the expressions. **Step 2** Change each expression into an equivalent expression with the LCD as the denominator. **Step 3** Add just as with expressions with like denominators. **Step 4** Simplify if necessary.

Example 1 Find $\dfrac{n+3}{n} + \dfrac{8n-4}{4n}$.

Factor each denominator.
$n = n$
$4n = 4 \cdot n$
LCD = $4n$

Since the denominator of $\dfrac{8n-4}{4n}$ is already $4n$, only $\dfrac{n+3}{n}$ needs to be renamed.

$\dfrac{n+3}{n} + \dfrac{8n-4}{4n} = \dfrac{4(n+3)}{4n} + \dfrac{8n-4}{4n}$
$= \dfrac{4n+12}{4n} + \dfrac{8n-4}{4n}$
$= \dfrac{12n+8}{4n}$
$= \dfrac{3n+2}{n}$

Example 2 Find $\dfrac{1}{2x^2+6x} + \dfrac{3}{x^2}$.

$\dfrac{1}{2x^2+6x} + \dfrac{3}{x^2} = \dfrac{1}{2x(x+3)} + \dfrac{3}{x^2}$
$= \dfrac{1}{2x(x+3)} \cdot \dfrac{x}{x} + \dfrac{3}{x^2} \cdot \dfrac{2(x+3)}{2(x+3)}$
$= \dfrac{x}{2x^2(x+3)} + \dfrac{6(x+3)}{2x^2(x+3)}$
$= \dfrac{x+6x+18}{2x^2(x+3)}$
$= \dfrac{7x+18}{2x^2(x+3)}$

Exercises

Find each sum.

1. $\dfrac{1}{a} + \dfrac{7}{3a}$

2. $\dfrac{1}{6x} + \dfrac{3}{8}$

3. $\dfrac{4}{9x} + \dfrac{5}{x^2}$

4. $\dfrac{2}{x^2} + \dfrac{3}{x^3}$

5. $\dfrac{8}{4a^2} + \dfrac{6}{3a}$

6. $\dfrac{4}{h+1} + \dfrac{2}{h+2}$

7. $\dfrac{4}{y+6} + \dfrac{1}{y+2}$

8. $\dfrac{y}{y^2+4y+4} + \dfrac{2}{y+2}$

9. $\dfrac{a}{a+4} + \dfrac{4}{a-4}$

10. $\dfrac{6}{3(m+1)} + \dfrac{2}{3(m-1)}$

11. $\dfrac{4x}{6x-2y} + \dfrac{3y}{9x-3y}$

12. $\dfrac{a-2}{a^2-4} + \dfrac{a-2}{a+2}$

13. $\dfrac{y+2}{y^2+5y+6} + \dfrac{2-y}{y^2+y-6}$

14. $\dfrac{q}{q^2-16} + \dfrac{q+1}{q^2+5q+4}$

Study Guide and Intervention 149 Glencoe Algebra 1

NAME _____ DATE _____ PERIOD _____

11-7 Study Guide and Intervention (continued)

Rational Expressions with Unlike Denominators

Subtract Rational Expressions Adapt the steps given on page 49 for adding rational expressions. In Step 3, subtract the numerators instead of adding them.

Example Find $\dfrac{3x}{x^2 - 4x} - \dfrac{1}{x - 4}$.

$$\dfrac{3x}{x^2 - 4x} - \dfrac{1}{x - 4} = \dfrac{3x}{x(x - 4)} - \dfrac{1}{x - 4} \quad \text{Factor the denominator.}$$

$$= \dfrac{3x}{x(x - 4)} - \dfrac{1}{x - 4} \cdot \dfrac{x}{x} \quad \text{The LCD is } x(x - 4).$$

$$= \dfrac{3x}{x(x - 4)} - \dfrac{x}{x(x - 4)} \quad 1 \cdot x = x$$

$$= \dfrac{2x}{x(x - 4)} \quad \text{Subtract numerators.}$$

$$= \dfrac{2}{x - 4} \quad \text{Simplify.}$$

Exercises

Find each difference.

1. $\dfrac{1}{a} - \dfrac{9}{4a}$

2. $\dfrac{1}{9x} - \dfrac{1}{8}$

3. $\dfrac{5}{9x} - \dfrac{1}{x^2}$

4. $\dfrac{6}{x^2} - \dfrac{3}{x^3}$

5. $\dfrac{5}{4a^2} - \dfrac{2}{3a}$

6. $\dfrac{h}{6h + 6} - \dfrac{1}{h + 1}$

7. $\dfrac{y}{y - 3} - \dfrac{3}{y + 3}$

8. $\dfrac{y}{y - 7} - \dfrac{y + 3}{y^2 - 4y - 21}$

9. $\dfrac{7a + 4}{3a + 9} - \dfrac{2a}{a + 3}$

10. $\dfrac{5}{m + 1} - \dfrac{2}{3(m - 1)}$

11. $\dfrac{4}{x - 2y} - \dfrac{2}{x + 2y}$

12. $\dfrac{a - 6b}{2a^2 - 5ab + 2b^2} - \dfrac{7}{a - 2b}$

13. $\dfrac{2}{y^2 + 3y + 2} - \dfrac{4}{y^2 + 2y + 1}$

14. $\dfrac{q}{q^2 + 2q + 1} - \dfrac{1}{q^2 + 5q + 4}$

NAME _____ DATE _____ PERIOD _____

11-8 Study Guide and Intervention

Mixed Expressions and Complex Fractions

Simplify Mixed Expressions Algebraic expressions such as $a + \dfrac{b}{c}$ and $5 + \dfrac{x+y}{x+3}$ are called **mixed expressions**. Changing mixed expressions to rational expressions is similar to changing mixed numbers to improper fractions.

Example 1 Simplify $5 + \dfrac{2}{n}$.

$5 + \dfrac{2}{n} = \dfrac{5 \cdot n}{n} + \dfrac{2}{n}$ LCD is n.

$= \dfrac{5n + 2}{n}$ Add the numerators.

Therefore, $5 + \dfrac{2}{n} = \dfrac{5n+2}{n}$.

Example 2 Simplify $2 + \dfrac{3}{n+3}$.

$2 + \dfrac{3}{n+3} = \dfrac{2(n+3)}{n+3} + \dfrac{3}{n+3}$

$= \dfrac{2n+6}{n+3} + \dfrac{3}{n+3}$

$= \dfrac{2n+6+3}{n+3}$

$= \dfrac{2n+9}{n+3}$

Therefore, $2 + \dfrac{3}{n+3} = \dfrac{2n+9}{n+3}$.

Exercises

Write each mixed expression as a rational expression.

1. $4 + \dfrac{6}{a}$

2. $\dfrac{1}{9x} - 3$

3. $3x - \dfrac{1}{x^2}$

4. $\dfrac{4}{x^2} - 2$

5. $10 + \dfrac{60}{x+5}$

6. $\dfrac{h}{h+4} + 2$

7. $\dfrac{y}{y-2} + y^2$

8. $4 - \dfrac{4}{2x+1}$

9. $1 + \dfrac{1}{x}$

10. $\dfrac{4}{m-2} - 2m$

11. $x^2 + \dfrac{x+2}{x-3}$

12. $a - 3 + \dfrac{a-2}{a+3}$

13. $4m + \dfrac{3n}{2t}$

14. $2q^2 + \dfrac{q}{p+q}$

15. $\dfrac{2}{y^2-1} - 4y^2$

16. $q^2 + \dfrac{p+q}{p-q}$

Study Guide and Intervention Glencoe Algebra 1

11-8 Study Guide and Intervention (continued)

Mixed Expressions and Complex Fractions

Simplify Complex Fractions If a fraction has one or more fractions in the numerator or denominator, it is called a **complex fraction**.

Simplifying a Complex Fraction	Any complex fraction $\dfrac{\frac{a}{b}}{\frac{c}{b}}$ where $b \neq 0$, $c \neq 0$, and $d \neq 0$, can be expressed as $\dfrac{ad}{bc}$.

Example Simplify $\dfrac{2 + \frac{4}{a}}{\frac{a+2}{3}}$.

$\dfrac{2 + \frac{4}{a}}{\frac{a+2}{3}} = \dfrac{\frac{2a}{a} + \frac{4}{a}}{\frac{a+2}{3}}$ Find the LCD for the numerator and rewrite as like fractions.

$= \dfrac{\frac{2a+4}{a}}{\frac{a+2}{3}}$ Simplify the numerator.

$= \dfrac{2a+4}{a} \cdot \dfrac{3}{a+2}$ Rewrite as the product of the numerator and the reciprocal of the denominator.

$= \dfrac{2(a+2)}{a} \cdot \dfrac{3}{a+2}$ Factor.

$= \dfrac{6}{a}$ Divide and simplify.

Exercises

Simplify each expression.

1. $\dfrac{2\frac{2}{5}}{3\frac{3}{4}}$

2. $\dfrac{\frac{3}{x}}{\frac{4}{y}}$

3. $\dfrac{\frac{x}{y^3}}{\frac{x^3}{y^2}}$

4. $\dfrac{1 - \frac{1}{x}}{1 + \frac{1}{x}}$

5. $\dfrac{1 - \frac{1}{x}}{1 - \frac{1}{x^2}}$

6. $\dfrac{\frac{1}{x-3}}{\frac{2}{x^2-9}}$

7. $\dfrac{\frac{x^2-25}{y}}{x^3 - 5x^2}$

8. $\dfrac{x - \frac{12}{x-1}}{x - \frac{8}{x-2}}$

9. $\dfrac{\frac{3}{y+2} - \frac{2}{y-2}}{\frac{1}{y+2} - \frac{2}{y-2}}$

NAME _____ **DATE** _____ **PERIOD** _____

11-9 Study Guide and Intervention

Solving Rational Equations

Solve Rational Equations Rational equations are equations that contain rational expressions. To solve equations containing rational expressions, multiply each side of the equation by the least common denominator.

Rational equations can be used to solve **work problems** and **rate problems**.

Example 1 Solve $\dfrac{x-3}{3} + \dfrac{x}{2} = 4$.

$\dfrac{x-3}{3} + \dfrac{x}{2} = 4$

$6\left(\dfrac{x-3}{3} + \dfrac{x}{2}\right) = 6(4)$ The LCD is 6.

$2(x-3) + 3x = 24$ Distributive Property

$2x - 6 + 3x = 24$ Distributive Property

$5x = 30$ Simplify.

$x = 6$ Divide each side by 5.

The solution is 6.

Example 2 **WORK PROBLEM** Marla can paint Percy's kitchen in 3 hours. Percy can paint it in 2 hours. Working together, how long will it take Marla and Percy to paint the kitchen?

In t hours, Marla completes $t \cdot \dfrac{1}{3}$ of the job and Percy completes $t \cdot \dfrac{1}{2}$ of the job. So an equation for completing the whole job is $\dfrac{t}{3} + \dfrac{t}{2} = 1$.

$\dfrac{t}{3} + \dfrac{t}{2} = 1$

$2t + 3t = 6$ Multiply each term by 6.

$5t = 6$ Add like terms.

$t = \dfrac{6}{5}$ Solve.

So it will take Marla and Percy $1\dfrac{1}{5}$ hours to paint the room if they work together.

Exercises

Solve each equation.

1. $\dfrac{x-5}{5} + \dfrac{x}{4} = 8$

2. $\dfrac{3}{x} = \dfrac{6}{x+1}$

3. $\dfrac{x-1}{5} = \dfrac{2x-2}{15}$

4. $\dfrac{8}{n-1} = \dfrac{10}{n+1}$

5. $s - \dfrac{4}{s+3} = s + 3$

6. $\dfrac{m+4}{m} + \dfrac{m}{3} = \dfrac{m}{3}$

7. $\dfrac{q+4}{q-1} + \dfrac{q}{q+1} = 2$

8. $\dfrac{5-2x}{2} - \dfrac{4x+3}{6} = \dfrac{7x+2}{6}$

9. $\dfrac{m+1}{m-1} - \dfrac{m}{1-m} = 1$

10. $\dfrac{x^2-9}{x-3} + x^2 = 9$

11. **GREETING CARDS** It takes Kenesha 45 minutes to prepare 20 greeting cards. It takes Paula 30 minutes to prepare the same number of cards. Working together at this rate, how long will it take them to prepare the cards?

12. **BOATING** A motorboat went upstream at 15 miles per hour and returned downstream at 20 miles per hour. How far did the boat travel one way if the round trip took 3.5 hours?

Study Guide and Intervention Glencoe Algebra 1

NAME _____ DATE _____ PERIOD ____

11-9 Study Guide and Intervention (continued)

Solving Rational Equations

Extraneous Solutions When you use cross multiplication or use the LCD of two rational expressions, you may get values for the variable that are not solutions of the original equation. Such values are called **extraneous solutions**.

Example Solve $\dfrac{15}{x^2 - 1} = \dfrac{5}{2(x - 1)}$.

$\dfrac{15}{x^2 - 1} = \dfrac{5}{2(x - 1)}$ Original equation

$30(x - 1) = 5(x^2 - 1)$ Cross multiply.

$30x - 30 = 5x^2 - 5$ Distributive Property

$0 = 5x^2 - 30x + 30 - 5$ Add $-30x + 30$ to each side.

$0 = 5x^2 - 30x + 25$ Simplify.

$0 = 5(x^2 - 6x + 5)$ Factor.

$0 = 5(x - 1)(x - 5)$ Factor.

$x = 1$ or $x = 5$ Zero Product Property

The number 1 is an extraneous solution, since 1 is an excluded value for x. So, 5 is the solution of the equation.

Exercises

Solve each equation. State any extraneous solutions.

1. $\dfrac{6x}{x - 1} + \dfrac{2x - 8}{x - 1} = 4$

2. $\dfrac{4x}{-x - 2} = \dfrac{1}{x + 2}$

3. $\dfrac{-5}{x + 2} = \dfrac{x - 1}{3}$

4. $\dfrac{x}{x - 3} + \dfrac{4}{3 - x} = x$

5. $\dfrac{x}{x - 2} - \dfrac{4}{2 - x} = x$

6. $\dfrac{x}{x^2 - 25} = \dfrac{1}{x + 5}$

7. $\dfrac{2}{x^2 - 36} - \dfrac{1}{x - 6} = 0$

8. $\dfrac{4z}{z^2 + 4z + 3} = \dfrac{6}{z + 3} + \dfrac{4}{z + 1}$

9. $\dfrac{4}{4 - p} - \dfrac{p^2}{p - 4} = 4$

10. $\dfrac{x^2 - 16}{x - 4} + x^2 = 16$

12-1 Study Guide and Intervention
Sampling and Bias

Sampling Techniques Suppose you want to survey students about their choice of radio stations. All students make up the **population** you want to survey. A **sample** is some portion of the larger group that you select to represent the entire group. A **census** would include all students within the population. A **random sample** of a population is selected so that it is representative of the entire population.

Simple Random Sample	a sample that is as likely to be chosen as another from a population
Stratified Random Sample	A population is first divided into similar, nonoverlapping groups. A simple random sample is then chosen from each group.
Systematic Random Sample	Items are selected according to a specified time or interval.

Example 1 SCHOOL Ten students are chosen randomly from each high school class to be on an advisory committee with the principal.

a. **Identify the sample and suggest a population from which it was chosen.**

The sample is 4 groups of 10 students each from the freshmen, sophomore, junior, and senior classes. The population is the entire student body of the school.

b. **Classify the sample as *simple*, *stratified*, or *systematic*.**

This is a stratified random sample because the population was first divided into nonoverlapping groups and then a random sample was chosen from each group.

Example 2 DOOR PRIZES Each of the participants in a conference was given a numbered name tag. Twenty-five numbers were chosen at random to receive a door prize.

a. **Identify the sample and suggest a population from which it was chosen.**

The sample was 25 participants of the conference. The population was all of the participants of the conference.

b. **Classify the sample as *simple*, *stratified*, or *systematic*.**

Since the numbers were chosen randomly, this is a simple random sample because each participant was equally likely to be chosen.

Exercises

Identify each sample, suggest a population from which it was selected, and classify the sample as *simple*, *stratified*, or *systematic*.

1. **SCHOOL** Each student in a class of 25 students was given a number at the beginning of the year. Periodically, the teacher chooses 4 numbers at random to display their homework on the overhead projector.

2. **GARDENING** A gardener divided a lot into 25-square-foot sections. He then took 2 soil samples from each and tested the samples for mineral content.

3. **SCHOOL** One hundred students in the lunch room are chosen for a survey. All students in the school eat lunch at the same time.

4. **SHOPPING** Every tenth person leaving a grocery store was asked if they would participate in a community survey.

NAME _____ DATE _____ PERIOD _____

12-1 Study Guide and Intervention (continued)
Sampling and Bias

Biased Sample A **biased sample** is one in which one or more parts of the population is favored over the other. Random samples are unbiased because each unit is selected without favoritism.

Biased samples include **convenience samples** in which members of the population are included because they are the most convenient to choose. A **voluntary response sample** is another type of biased sample that includes only those members of the population who choose to participate in the sampling.

Example SCHOOL The principal of a high school wanted to know if students in the school liked the attendance policy. He decided to survey the students in the third-hour study hall about whether they like the attendance policy. Fewer than one fourth of the students in the school have a study hall.

a. **Identify the sample and suggest a population from which it is chosen.**

The sample includes only those students in the third-hour study hall. The population is the entire student body.

b. **Classify the sample as *convenience* or *voluntary response*.**

This sample is a convenience sample because it is convenient to sample students in a study hall during a certain period of the day.

Exercises

Identify each sample, suggest a population from which it was selected, and classify the sample as *convenience* or *voluntary response*.

1. **SCHOOL** The high school administration wanted to evaluate how homecoming week was conducted at the school. Each female member of the Student Council at the high school was asked if she liked homecoming activities.

2. **MANUFACTURING** A clothing company wanted to check quality control of all its products. The plant manager decided to look at every fourth item inspected by Inspector X. There are 10 inspectors in the plant.

3. **SCHOOL** The counselors of a high school sent out a survey to senior students with questions about their plans for college. Some students did not plan to attend college. 40% of the seniors sent responses back.

4. **BUSINESS** A marketing group was asked to compile data on the effectiveness of advertisements for household products across the country. The group chose to conduct surveys at shopping malls. Every person walking by a survey taker in a shopping mall was asked if he/she would participate in a survey about household products.

NAME _____ DATE _____ PERIOD _____

12-2 Study Guide and Intervention

Counting Outcomes

Tree Diagrams One method used for counting the number of possible outcomes of an event is to draw a **tree diagram**. The last column of the tree diagram shows all of the possible outcomes. The list of all possible outcomes is called the **sample space**, and a specific outcome is called an **event**.

Example 1 Suppose you can set up a stereo system with a choice of video, DVD, or laser disk players, a choice of cassette or graphic equalizer audio components, and a choice of single or dual speakers. Draw a tree diagram to show the sample space.

Player	Audio	Speaker	Outcomes
video	cassette	Single / Dual	VCS / VCD
	graphic equalizer	Single / Dual	VGS / VGD
DVD	cassette	Single / Dual	DCS / DCD
	graphic equalizer	Single / Dual	DGS / DGD
laser disk	cassette	Single / Dual	LCS / LCD
	graphic equalizer	Single / Dual	LGS / LGD

The tree diagram shows that there are 12 ways to set up the stereo system.

Example 2 A food stand offers ice cream cones in vanilla or chocolate flavors. It also offers fudge or caramel toppings, and it uses sugar or cake cones. Use a tree diagram to determine the number of possible ice cream cones.

Flavor	Toppings	Cone	Outcomes
vanilla	fudge	sugar / cake	VFS / VFC
	caramel	sugar / cake	VCS / VCC
chocolate	fudge	sugar / cake	CFS / CFC
	caramel	sugar / cake	CCS / CCC

The tree diagram shows that there are 8 possible ice cream cones.

Exercises

The spinner at the right is spun twice.

1. Draw a tree diagram to show the sample space.

2. How many outcomes are possible?

A pizza can be ordered with a choice of sausage, pepperoni, or mushrooms for toppings, a choice of thin or pan for the crust, and a choice of medium or large for the size.

3. Draw a tree diagram to show the sample space.

4. How many pizzas are possible?

Study Guide and Intervention 157 Glencoe Algebra 1

12-2 Study Guide and Intervention (continued)

Counting Outcomes

The Fundamental Counting Principle Another way to count the number of possible outcomes is to use the Fundamental Counting Principle.

Fundamental Counting Principle	If an event M can occur in m ways and an event N can occur in n ways, then M followed by N can occur in $m \cdot n$ ways.

Example
Carly and Jake went to an arcade with 9 different games.

a. In how many different orders can they play the games if they play each one only once?

The number of orders for playing can be found by multiplying the number of choices for each position. Let n represent the number of possible orders.

$n = 9 \cdot 8 \cdot 7 \cdot 6 \cdot 5 \cdot 4 \cdot 3 \cdot 2 \cdot 1 = 362,880$

There are 362,880 ways to play each of 9 arcade games once. This is also known as a **factorial**, or $n = 9! = 9 \cdot 8 \cdot 7 \cdot 6 \cdot 5 \cdot 4 \cdot 3 \cdot 2 \cdot 1$.

b. If they have only enough tokens to play 6 different games, how many ways can they do this?

Use the Fundamental Counting Principle to find the sample space. There are 9 choices for the first game, 8 choices for the second, and so on, down to 4 choices for the sixth game.

$n = 9 \cdot 8 \cdot 7 \cdot 6 \cdot 5 \cdot 4 = 60,480$

There are 60,480 ways to play 6 different arcade games once.

Exercises

Find the value of each expression.

1. 6!

2. 11!

3. 8!

4. A sub sandwich restaurant offers four types of sub sandwiches, three different types of potato chips, five types of bread, and six different beverages. How many different sandwich and drink combinations can you order?

5. How many outfits are possible if you can choose one from each of four shirts, three pairs of pants, two pairs of shoes, and two jackets?

6. In how many ways can you arrange 5 boxes of cereal on a shelf?

7. Seven students sit in a row in the auditorium. In how many ways can they arrange themselves?

8. Kinjal puts 12 different books on a shelf. In how many different ways can she arrange them?

NAME _____ DATE _____ PERIOD _____

12-3 Study Guide and Intervention

Permutations and Combinations

Permutations An arrangement or listing in which order or placement is important is called a **permutation**. For example the arrangement AB of choices A and B is different from the arrangement BA of these same two choices.

Permutations	$_nP_r = \dfrac{n!}{(n-r)!}$

Example 1 Find $_6P_2$.

$_nP_r = \dfrac{n!}{(n-r)!}$ Definition of $_nP_r$

$_6P_2 = \dfrac{6!}{(6-2)!}$ $n = 6, r = 2$

$= \dfrac{6!}{4!}$ Simplify.

$= \dfrac{6 \cdot 5 \cdot 4 \cdot 3 \cdot 2 \cdot 1}{4 \cdot 3 \cdot 2 \cdot 1}$ Definition of factorial

$= 6 \cdot 5$ or 30   Simplify.

There are 30 permutations of 6 objects taken 2 at a time.

Example 2 A specific program requires the user to enter a 5-digit password. The digits cannot repeat and can be any five of the digits 1, 2, 3, 4, 7, 8, and 9.

a. How many different passwords are possible?

$_nP_r = \dfrac{n!}{(n-r)!}$

$_7P_5 = \dfrac{7!}{(7-5)!}$

$= \dfrac{7 \cdot 6 \cdot 5 \cdot 4 \cdot 3 \cdot 2 \cdot 1}{2 \cdot 1}$

$= 7 \cdot 6 \cdot 5 \cdot 4 \cdot 3$ or 2520

There are 2520 ways to create a password.

b. What is the probability that the first two digits are odd numbers with the other digits any of the remaining numbers?

$P(\text{first two digits odd}) = \dfrac{\text{number of favorable outcomes}}{\text{number of possible outcomes}}$

Since there are 4 odd digits, the number of choices for the first digit is 4, and the number of choices for the second digit is 3. Then there are 5 choices left for the third digit, 4 for the fourth, and 3 for the fifth, so the number of favorable outcomes is $4 \cdot 3 \cdot 5 \cdot 4 \cdot 3$, or 720.

The probability is $\dfrac{720}{2520} \approx 28.6\%$.

Exercises

Evaluate each expression.

1. $_7P_4$

2. $_{12}P_7$

3. $(_9P_9)(_{16}P_2)$

4. A club with ten members wants to choose a president, vice-president, secretary, and treasurer. Six of the members are women, and four are men.

   a. How many different sets of officers are possible?

   b. What is the probability that all officers will be women.

Study Guide and Intervention   Glencoe Algebra 1

12-3 Study Guide and Intervention (continued)

Permutations and Combinations

Combinations An arrangement or listing in which order is not important is called a **combination**. For example, AB and BA are the same combination of A and B.

Combinations	$_nC_r = \dfrac{n!}{(n-r)!r!}$

Example A club with ten members wants to choose a committee of four members. Six of the members are women, and four are men.

a. How many different committees are possible?

$_nC_r = \dfrac{n!}{(n-r)!r!}$ Definition of combination

$= \dfrac{10!}{(10-4)!4!}$ $n = 10$, $r = 4$

$= \dfrac{10 \cdot 9 \cdot 8 \cdot 7}{4!}$ Divide by the GCF 6!.

$= 210$ Simplify.

There are 210 ways to choose a committee of four when order is not important.

b. If the committee is chosen randomly, what is the probability that two members of the committee are men?

There are $_4C_2 = \dfrac{4!}{(4-2)!2!} = 6$ ways to choose two men randomly, and there are $_6C_2 = \dfrac{6!}{(6-4)!4!} = 15$ ways to choose two women randomly. By the Fundamental Counting Principle, there are $6 \cdot 15$ or 90 ways to choose a committee with two men and two women.

Probability (2 men and 2 women) $= \dfrac{\text{number of favorable outcomes}}{\text{number of possible outcomes}}$

$= \dfrac{90}{210}$ or about 42.9%

Exercises

Find the value of each expression.

1. $_7C_3$

2. $_{12}C_8$

3. $(_9C_9)(_{11}C_9)$

4. In how many ways can a club with 9 members choose a two-member sub-committee?

5. A book club offers its members a book each month for a year from a selection of 24 books. Ten of the books are biographies and 14 of the books are fiction.

   a. How many ways could the members select 12 books?

   b. What is the probability that 5 biographies and 7 fiction books will be chosen?

12-4 Study Guide and Intervention

Probability of Compound Events

Independent and Dependent Events Compound events are made up of two or more simple events. The events can be **independent events** or they can be **dependent events**.

Probability of Independent Events	Outcome of first event does not affect outcome of second.	$P(A \text{ and } B) = P(A) \cdot P(B)$	Example: rolling a 6 on a die and then rolling a 5
Probability of Dependent Events	Outcome of first event does affect outcome of second.	$P(A \text{ and } B) = P(A) \cdot P(B \text{ following } A)$	Example: without replacing the first card, choosing an ace and then a king from a deck of cards

Example 1 Find the probability that you will roll a six and then a five when you roll a die twice.

By the definition of independent events, $P(A \text{ and } B) = P(A) \cdot P(B)$

First roll: $P(6) = \frac{1}{6}$

Second roll: $P(5) = \frac{1}{6}$

$P(6 \text{ and } 5) = P(6) \cdot P(5)$
$= \frac{1}{6} \cdot \frac{1}{6}$
$= \frac{1}{36}$

The probability that you will roll a six and then roll a five is $\frac{1}{36}$.

Example 2 A bag contains 3 red marbles, 2 green marbles, and 4 blue marbles. Two marbles are drawn randomly from the bag and not replaced. Find the probability that both marbles are blue.

By the definition of dependent events, $P(A \text{ and } B) = P(A) \cdot P(B \text{ following } A)$

First marble: $P(\text{blue}) = \frac{4}{9}$

Second marble: $P(\text{blue}) = \frac{3}{8}$

$P(\text{blue, blue}) = \frac{4}{9} \cdot \frac{3}{8}$
$= \frac{12}{72}$
$= \frac{1}{6}$

The probability of drawing two blue marbles is $\frac{1}{6}$.

Exercises

A bag contains 3 red, 4 blue, and 6 yellow marbles. One marble is selected at a time, and once a marble is selected, it is not replaced. Find each probability.

1. $P(2 \text{ yellow})$

2. $P(\text{red, yellow})$

3. $P(\text{blue, red, yellow})$

4. George has two red socks and two white socks in a drawer. What is the probability of picking a red sock and a white sock in that order if the first sock is not replaced?

5. Phyllis drops a penny in a pond, and then she drops a nickel in the pond. What is the probability that both coins land with tails showing?

6. A die is rolled and a penny is dropped. Find the probability of rolling a two and showing a tail.

NAME _____ DATE _____ PERIOD _____

12-4 Study Guide and Intervention (continued)

Probability of Compound Events

Mutually Exclusive and Inclusive Events Events that cannot occur at the same time are called **mutually exclusive**. If two events are not mutually exclusive, they are called **inclusive**.

Probability of Mutually Exclusive Events	$P(A \text{ or } B) = P(A) + P(B)$	$P(\text{rolling a 2 or a 3 on a die}) = P(2) + P(3) = \frac{1}{3}$
Probability of Inclusive Events	$P(A \text{ or } B) = P(A) + P(B) - P(A \text{ and } B)$	$P(\text{King or Heart}) = P(K) + P(H) - P(K \text{ and } H) = \frac{9}{26}$

Example 1 One card is drawn from a standard deck of 52 cards. Find the probability of drawing a king or a queen.

Drawing a king or a queen are mutually exclusive events.
By the definition of mutually exclusive events, $P(A \text{ or } B) = P(A) + P(B)$.

$P(A) = P(\text{king}) = \frac{4}{52} = \frac{1}{13}$ $P(B) = P(\text{queen}) = \frac{4}{52} = \frac{1}{13}$

$P(\text{king or queen}) = \frac{1}{13} + \frac{1}{13}$

$= \frac{2}{13}$

The probability of drawing a king or a queen is $\frac{2}{13}$.

Exercises

A bag contains 2 red, 5 blue, and 7 yellow marbles. Find each probability.

1. $P(\text{yellow or red})$
2. $P(\text{red or not yellow})$
3. $P(\text{blue or red or yellow})$

One card is drawn from a standard deck of 52 cards. Find each probability.

4. $P(\text{jack or red})$
5. $P(\text{red or black})$

6. $P(\text{jack or clubs})$
7. $P(\text{queen or less than 3})$

8. $P(5 \text{ or } 6)$
9. $P(\text{diamond or spade})$

10. In a math class, 12 out of 15 girls are 14 years old and 14 out of 17 boys are 14 years old. What is the probability of selecting a girl or a 14-year old from this class?

12-5 Study Guide and Intervention

Probability Distributions

Random Variables and Probability Distributions A random variable X is a variable whose value is the numerical outcome of a random event.

Example A teacher asked her students how many siblings they have. The results are shown in the table at the right.

Number of Siblings	Number of Students
0	1
1	15
2	8
3	2
4	1

a. **Find the probability that a randomly selected student has 2 siblings.**

The random variable X can equal 0, 1, 2, 3, or 4. In the table, the value $X = 2$ is paired with 8 outcomes, and there are 27 students surveyed.

$$P(X = 2) = \frac{2 \text{ siblings}}{27 \text{ students surveyed}}$$

$$= \frac{8}{27}$$

The probability that a randomly selected student has 2 siblings is $\frac{8}{27}$, or 29.6%.

b. **Find the probability that a randomly selected student has at least three siblings.**

$$P(X \geq 3) = \frac{2 + 1}{27}$$

The probability that a randomly selected student has at least 3 siblings is $\frac{1}{9}$, or 11.1%.

Exercises

For Exercises 1–3, use the grade distribution shown at the right. A grade of A = 5, B = 4, C = 3, D = 2, F = 1.

X = Grade	5	4	3	2	1
Number of Students	6	9	5	4	1

1. Find the probability that a randomly selected student in this class received a grade of C.

2. Find the probability that a randomly selected student in this class received a grade lower than a C.

3. What is the probability that a randomly selected student in this class passes the course, that is, gets at least a D?

4. The table shows the results of tossing 3 coins 50 times. What is the probability of getting 2 or 3 heads?

X = Number of Heads	0	1	2	3
Number of Times	6	20	19	5

Study Guide and Intervention   Glencoe Algebra 1

12-5 Study Guide and Intervention (continued)
Probability Distributions

Probability Distributions The probabilities associated with every possible value of the random variable X make up what are called the **probability distribution** for that variable. A probability distribution has the following properties.

Properties of a Probability Distribution	1. The probability of each value of X is greater than or equal to 0. 2. The probabilities for all values of X add up to 1.

The probability distribution for a random variable can be given in a table or in a **probability histogram** and used to obtain other information.

Example The data from the example on the previous page can be used to determine a probability distribution and to make a probability histogram.

X = Number of Siblings	P(X)
0	0.037
1	0.556
2	0.296
3	0.074
4	0.037

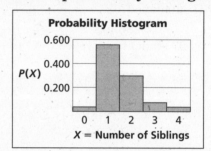

a. Show that the probability distribution is valid.

For each value of X, the probability is greater than or equal to 0 and less than or equal to 1. Also, the sum of the probabilities is 1.

b. What is the probability that a student chosen at random has fewer than 2 siblings?

Because the events are independent, the probability of fewer than 2 siblings is the sum of the probability of 0 siblings and the probability of 1 sibling, or $0.037 + 0.556 = 0.593$.

Exercises

The table at the right shows the probability distribution for students by school enrollment in the United States in 2000. Use the table for Exercises 1–3.

X = Type of School	P(X)
Elementary = 1	0.562
Secondary = 2	0.215
Higher Education = 3	0.223

Source: U.S. Census Bureau

1. Show that the probability distribution is valid.

2. If a student is chosen at random, what is the probability that the student is in elementary or secondary school?

3. Make a probability histogram of the data.

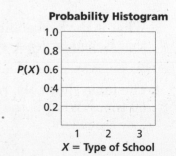

NAME _____ DATE _____ PERIOD _____

12-6 Study Guide and Intervention

Probability Simulations

Theoretical and Experimental Probability The probability used to describe events mathematically is called **theoretical probability**. For example, the mathematical probability of rolling a 4 with a number cube is $\frac{1}{6}$, or $P(4) = \frac{1}{6}$. **Experimental probability** is the ratio of the number of times an outcome occurs in an experiment to the total number of events or trials, known as the **relative frequency**.

Experimental probability	$\frac{\text{frequency of an outcome}}{\text{total number of trials}}$

Example 1 Matt recorded that it rained 8 times in November and snowed 3 times. The other days, it was sunny. There are 30 days in November. Suppose Matt uses these results to predict November's weather next year. What is the probability that a day in November will be sunny?

$$\text{Experimental Probability} = \frac{\text{frequency of outcome}}{\text{total number of trials}}$$

$$= \frac{(30 - 8 - 3)}{30}$$

$$= \frac{19}{30} = 63.3\%$$

The probability that it will be sunny on a day in November is 63.3%.

Example 2 A football team noticed that 9 of the last 20 coin tosses to choose which team would receive the ball first resulted in tails. What is the experimental probability of the coin landing on tails? What is the theoretical probability?

$$\text{Experimental Probability} = \frac{\text{frequency of outcome}}{\text{total number of trials}}$$

$$= \frac{\text{number of tails}}{\text{total number of tosses}}$$

$$= \frac{9}{20} = 45\%$$

In this case, the experimental probability that a coin toss will be tails is 45%. If the coin is fair, the mathematical probability is 50%.

Exercises

A math class decided to test whether a die is fair, that is, whether the experimental probability equals the theoretical probability. The results for 100 rolls are shown at the right. Use the information for Exercises 1–3.

1: 1	2: 15
3: 4	4: 13
5: 15	6: 42

1. What is the theoretical probability of rolling a 6?

2. What is the experimental probability of rolling a 6?

3. Is the die fair? Explain your reasoning.

Study Guide and Intervention   165   Glencoe Algebra 1

NAME _____ DATE _____ PERIOD _____

12-6 Study Guide and Intervention *(continued)*

Probability Simulations

Performing Simulations A method that is often used to find experimental probability is a **simulation**. A simulation allows you to use objects to act out an event that would be difficult or impractical to perform.

Example In one baseball season, Pete was able to get a base hit 42 of the 254 times he was at bat.

a. What could be used to simulate his getting a base hit?

First find the experimental probability.

Experimental Probability = $\dfrac{\text{frequency of outcome}}{\text{total number of trials}}$

$= \dfrac{42}{254}$ or 16.5%

Notice that the experimental probability is about $\dfrac{1}{6}$. Therefore use a spinner like the one at the right with 6 equally likely outcomes.

b. Describe a way to simulate his next 10 times at bat.

Let an outcome of 1 correspond to Pete's getting a base hit. Let all other outcomes correspond to his *not* getting a hit. Spin the spinner once to simulate a time at bat. Record the result and repeat this 9 more times.

Exercises

1. What could you use to simulate the outcome of guessing on a 20 question true-false test?

2. What could you use to simulate the outcome of guessing on a 20-question multiple choice test with 4 alternative answers labeled A, B, C, and D for each question?

For Exercises 3–4, use the following information.
Main Street Supermarket randomly gives each shopper a free two-liter bottle of cola during the Saturday shopping hours. The supermarket sells 6 different types of cola.

3. What could be used to perform a simulation of this situation?

4. How could you use this simulation to model the next 50 bottles of cola given out.

5. At a picnic, there were 2 peanut butter sandwiches, 2 chicken sandwiches, a tuna sandwich, and a turkey sandwich in a cooler. Describe a simulation that could be used to find the probability of randomly picking a certain sandwich from the cooler.

Study Guide and Intervention 166 Glencoe Algebra 1